BRICKWORK AND ASSOCIATED STUDIES
Volume 3

BRICKWORK AND ASSOCIATED STUDIES
Volume 3

H. Bailey

D. W. Hancock

Senior Lecturers,
Stockport College of Technology

First published 1979 by
THE MACMILLAN PRESS LTD
London and Basingstoke
Associated companies in Delhi Dublin
Hong Kong Johannesburg Lagos Melbourne
New York Singapore and Tokyo

Typeset in 10/11pt Theme by
STYLESET LIMITED
Salisbury · Wiltshire

Printed in Great Britain by A. Wheaton & Co. Ltd., Exeter

British Library Cataloguing in Publication Data

Bailey, H
    Brickwork and associated studies.
    Vol.3
    1. Building, Brick
    I. Title II. Hancock, D W
    693.2'1   TH1301

    ISBN 0-333-26906-3 Pbk

CONTENTS

PREFACE

This series of three volumes is designed to provide an introduction to the brickwork craft and the construction industry for craft apprentices and all students involved in building. All too often, new entrants to the construction industry are expected to have a knowledge of calculations, geometry, science and technology irrespective of their previous education. It is the authors' aim to provide a course of study which is not only easily understood but is also able to show the relationship that exists between technology and associated studies.

The construction industry recognises that the modern craftsman, while maintaining a very high standard of skills, must be capable of accepting change — in methods, techniques and materials. Therefore it will be necessary for apprentices to develop new skills related to the constant advancements in technology.

This third volume concludes the complete Craft Certificate course for the City & Guilds of London Institute, and includes the many other areas of work in which the craftsman is required to demonstrate his ability.

To become a highly skilled technician in the modern construction industry, the apprentice should recognise that physical skills must be complemented by technology, and that planned methods of construction must be used in all work situations.

The apprentice and young craftsman will be able to appreciate the diversity of the bricklayer's craft, and to relate his own abilities and ambitions to the immense scope offered by today's construction industry.

H. BAILEY
D. W. HANCOCK

ACKNOWLEDGEMENTS

The authors wish to acknowledge the assistance and cooperation of: The Stone Firms Ltd, for figure 1.28; The Clay Pipe Development Association Ltd, for figures 5.19, 5.20 and 5.21; S.G.B. Scaffolding (Great Britain) Ltd, for figures 6.5 to 6.20, 6.25 and 6.29; The Brick Development Association, for figure 6.26; Walter Somers (Materials Handling) Ltd, for figures 6.36, 6.37 and 6.38.

1
CLADDING

DEFINITION AND MATERIALS

Cladding is the term used to describe the envelope of facing material that encloses the building structure. Cladding is tied into the structure, but does not accept any form of loading; the loading of the cladding is transferred at specified positions on to the structural frame (figure 1.1).

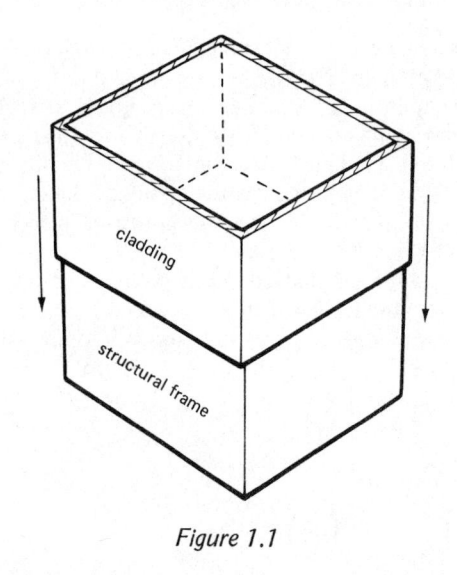

Figure 1.1

    The materials commonly used for cladding are stone (that is, sandstone and limestone), slate, granite, marble and concrete.

SITE WORK

Receiving Cladding Slabs on Site (figures 1.2 and 1.3)

When cladding slabs are delivered to a building site the unloading often causes problems because of the size and weight of the units and the nature of the material itself. There are three methods of unloading

(1) by lorry crane
(2) by site crane
(3) manual unloading.

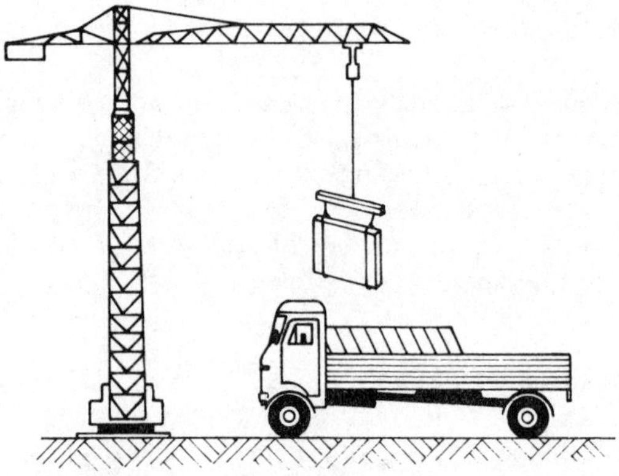

Figure 1.2

Lorry Crane

Unloading by lorry crane requires only the minimum amount of labour; slabs are lifted in packs or singly, depending on their size. When in packs the slabs are protected by spacing laths and secured in nylon slings. The slabs can be stacked and stored near where they are to be used, but obviously the effectiveness of this method depends on the transport facilities that are available on site.

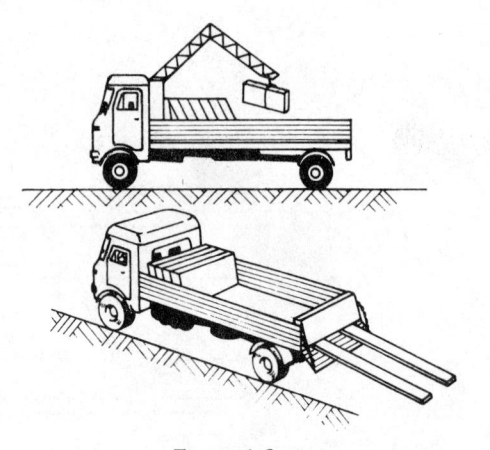

Figure 1.3

Site Crane

The advantage of this method is that slabs can be lifted from the lorry to the actual position where the fixing operation is being carried out. To use this method successfully it is essential to plan the delivery of the slabs to coincide with the programme for fixing, and to ensure that the site crane is available when required. This method reduces labour costs to a minimum.

Manual Unloading

This method is costly in labour, and often results in damage to the cladding slabs. The unloading is often carried out by 'wheeling off' by truck from the lorry. The storage position for the slabs is frequently determined by the amount of manual effort required and also site hazards.

Checking before Storage

It is essential to carry out a check on all slabs before stacking and storing them. The checking should involve

(1) examining any numbers or marks on the slabs to see that they coincide with the working drawings
(2) inspecting for surface defects, that is, fractures, abrasions or staining
(3) examining for squareness, dimensions and slotting positions.

The slabs are then stacked according to their order of use.

Stacking and Storage

When slabs are to be stored on site they should be positioned well clear of site buildings and equipment (figures 1.4 and 1.5). To prevent staining and to provide protection it is essential to stack the slabs on planks above ground level, using an A-frame timber support, and to ensure that all slabs are covered with plastic sheeting, which should be adequately secured.

Lifting on Site

The method adopted to lift cladding units is determined by

(1) the site equipment available
(2) the dimensions of the slabs.

Large slabs are either lifted singly or in pairs, depending on their size. The site crane is used and the slabs are placed in nylon slings or chains fixed to special fixing hooks inserted into the slabs (figures 1.6–1.10).

Medium and small slabs are often lifted by site crane in packs secured by nylon slings. Small slabs can also be lifted with the site hoist and placed on trucks (figures 1.7–1.9).

It is essential that adequate methods of protection are used when lifting operations are being carried out; this requires the use of compressible spacers and covering laths.

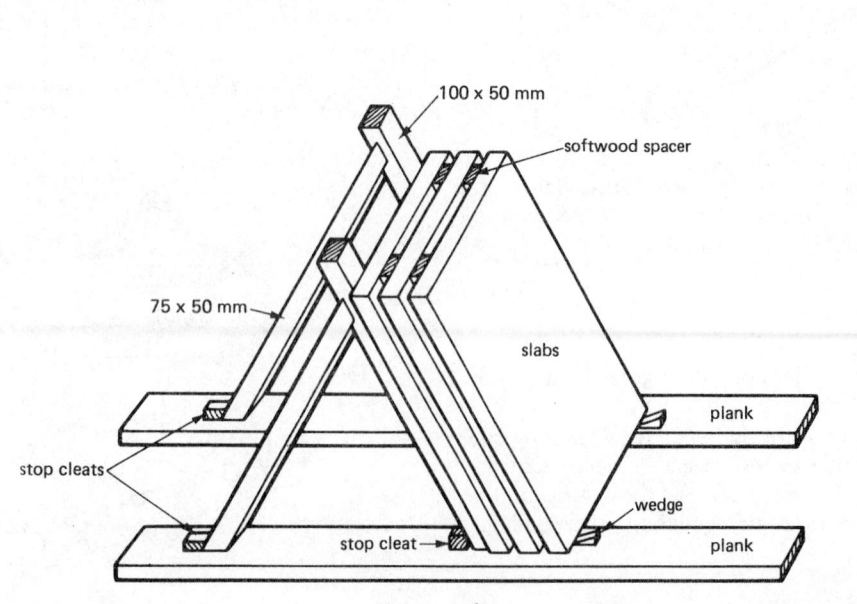

Figure 1.4

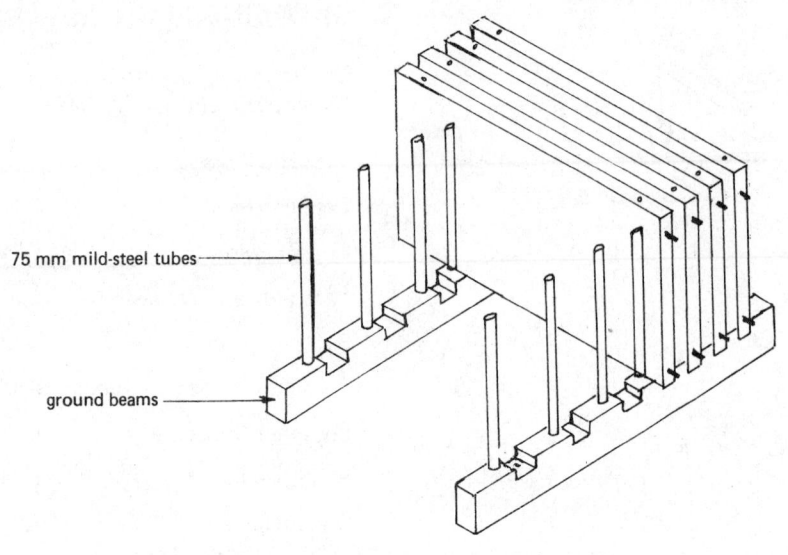

75 mm mild-steel tubes

ground beams

Figure 1.5 Stacking large slabs or panels on site

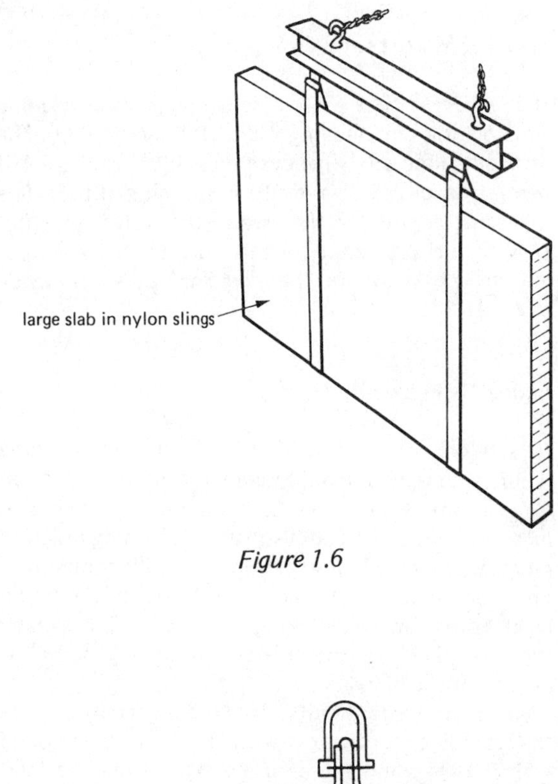

large slab in nylon slings

Figure 1.6

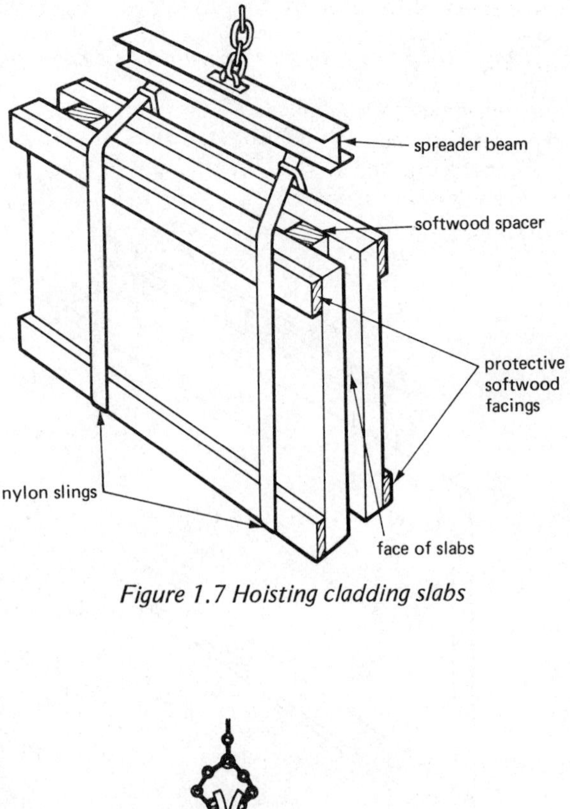

spreader beam

softwood spacer

protective
softwood
facings

nylon slings

face of slabs

Figure 1.7 Hoisting cladding slabs

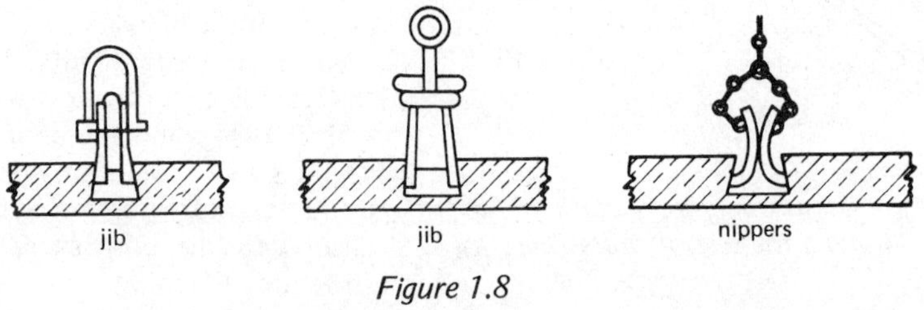

jib

jib

nippers

Figure 1.8

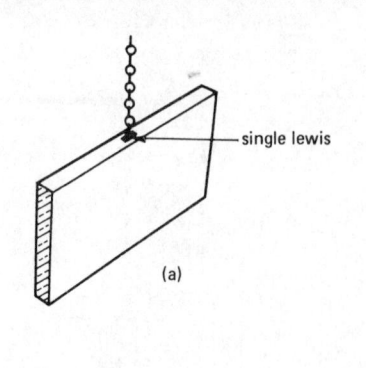

single lewis

(a)

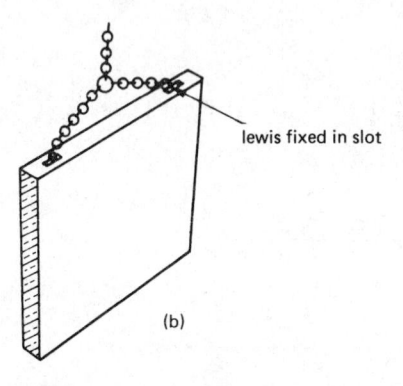

lewis fixed in slot

(b)

Figure 1.9 Lifting a single slab with (a) a single lewis and (b) a double lewis

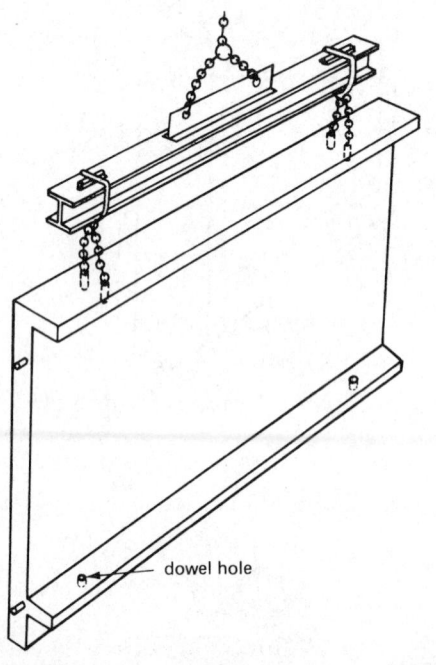

dowel hole

Figure 1.10 Lifting a large concrete cladding slab with a steel beam, showing the use of bolts fixed in screwed inserts

CLADDING FIXINGS

The following materials are recommended as fixings for all types of cladding slab

Material	Use
Copper, rolled	Cramps and anchors
Copper, wire rod	Dowels and cramps
Phosphor bronze, cast	Corbel plates and cramps
Phosphor bronze, rolled	Anchors and angles
Phosphor bronze, extruded	Anchors and Z-sections
Phosphor bronze, wire rod	Dowels and cramps
Stainless steel	All types
Gun metal	All types

All fixings should be of non-ferrous metal and should be made accurately, according to the designer's specifications. It is recommended that all fixings should be made from the same material whenever possible (figures 1.11 and 1.12).

Forming Mortices and Cramp Holes

It is essential that great care is taken when forming slots or holes in cladding slabs. Drilling should be the method adopted whenever possible. The use of percussion tools often results in stunning of the stone or even fracturing in the area of the slot or mortice. It is normal procedure for this work to be carried out at the works where the cladding slabs are made (figure 1.13).

Fixing Techniques

It is widely recognised that the fixing of cladding requires considerable skill, and there are firms specialising in this type of work; they employ their own masonry craftsmen and also supply cladding units. In modern building construction the skilled mason is now very much in demand, consequently the bricklayer craftsman is becoming involved in fixing many types of cladding, and this will ultimately be to the benefit of the industry.

It is the responsibility of the designer to provide the cladding contractor with 1:10, 1:20 and 1:5 scale drawings and also details of plans and elevations. The designer should provide the drawings so that the cladding contractor can supply the units in accordance with the programme for the erection of the structure. Cladding slabs are fixed to the concrete

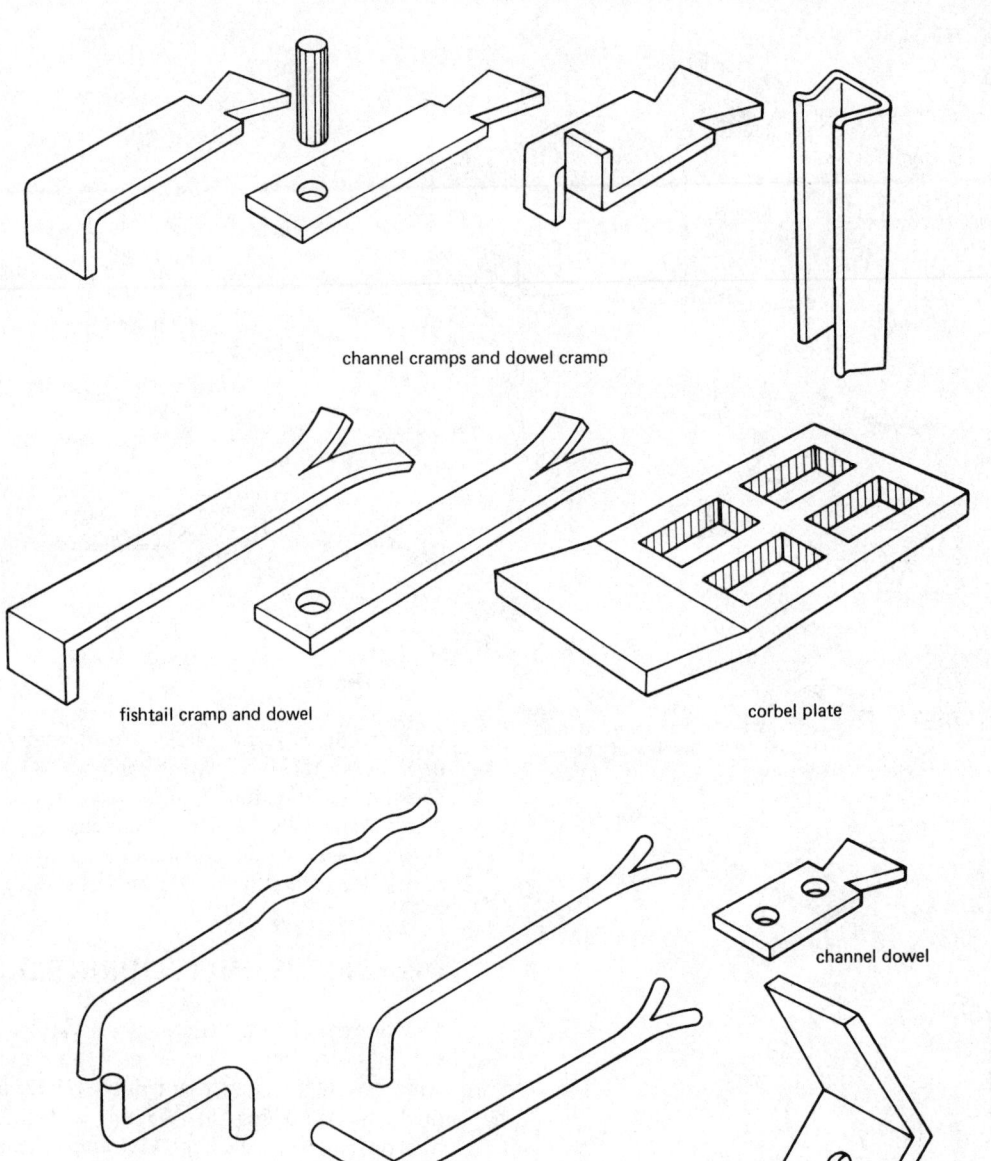

channel cramps and dowel cramp

fishtail cramp and dowel

corbel plate

rod cramps and hooks

channel dowel

channel corbel plate

Figure 1.11 Masonry anchors

structural frame as soon as the concrete is cured. When the backing is of brickwork, the cladding operations must be carried out at the same time (figures 1.14 and 1.15).

MORTAR FOR CLADDING

The type of mortar used for cladding varies according to the material being used, and also with the designer's

requirements. The general requirements for cladding mortar are

(1) the mortar should not be too dense or brittle, otherwise spalling will occur at the joints
(2) the mortar should be designed not to shrink when under pressure, or after setting action has been completed, because this could result in the penetration of water

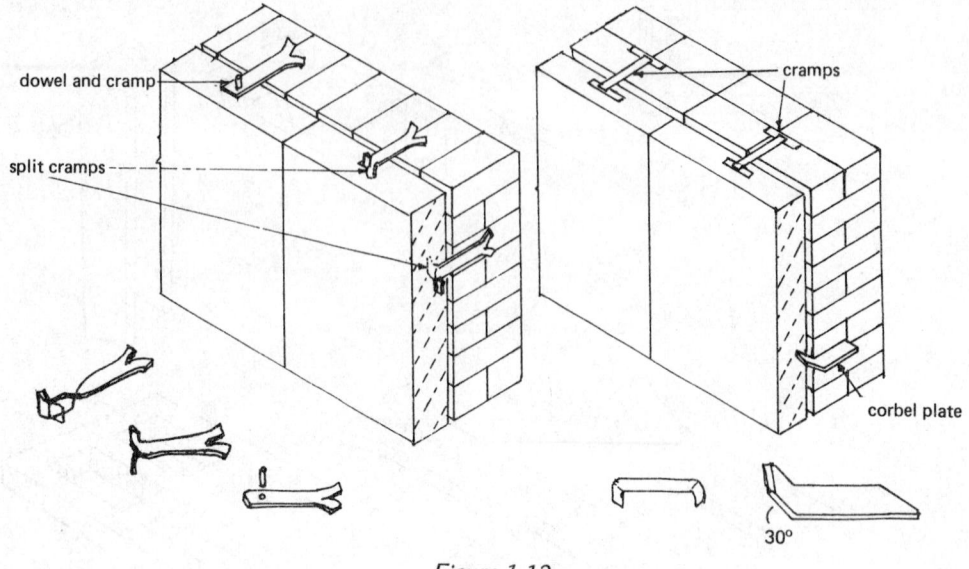

Figure 1.12

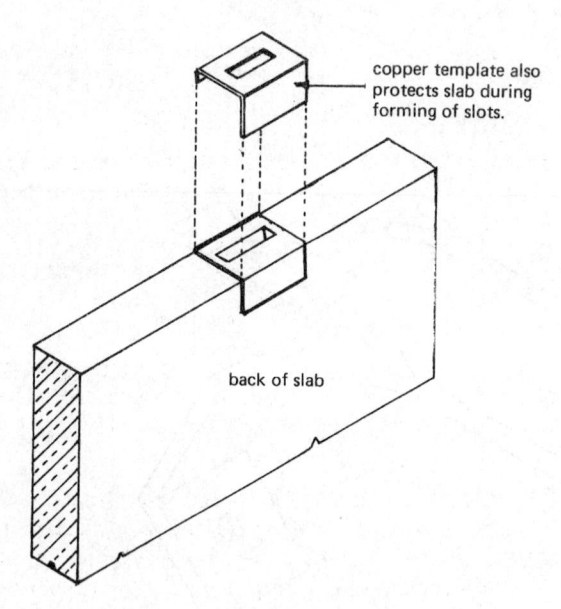

Figure 1.13

(3) the colour and surface texture should blend in with the cladding materials.

The general recommended mix for stone cladding is seven parts stone dust, five parts hydrated lime and one part white cement.

Joints and beds should not exceed 5–6 mm unless stated, with 12 mm grouting between concrete backing and cladding slab, when this is the designer's requirement.

Fixing Details

Cramps, dowels and hooks are normally used with brickwork backing. When concrete is used as the backing material the fixings used are usually corbel plates, channels, anchors, cramps and adjustable bolts. Projecting nibs are often incorporated into the concrete backing to provide vertical support.

DIMENSIONS OF CLADDING SLABS

The dimensions of cladding slabs vary according to the material used. For normal use sandstone and limestone slabs are recommended to have a minimum thickness of not less than 75 mm, with the maximum thickness not exceeding 100 mm. Obviously the increased thickness of slab will require fewer fixings and provide increased stability. Staining is also reduced by using slabs of 100 mm. The maximum recommended area for individual slabs is 1.02 m^2.

Granite, Marble and Slate Slabs

When used externally the recommended minimum thickness for these slabs is 32 mm. When used as internal facings, it is possible to reduce the thickness to 28 mm, with a maximum area not greater than 0.5 m^2.

Stone-faced Slabs

These can vary from 75 mm to 150 mm in thickness, with a recommended maximum area of 1.0 m^2. The

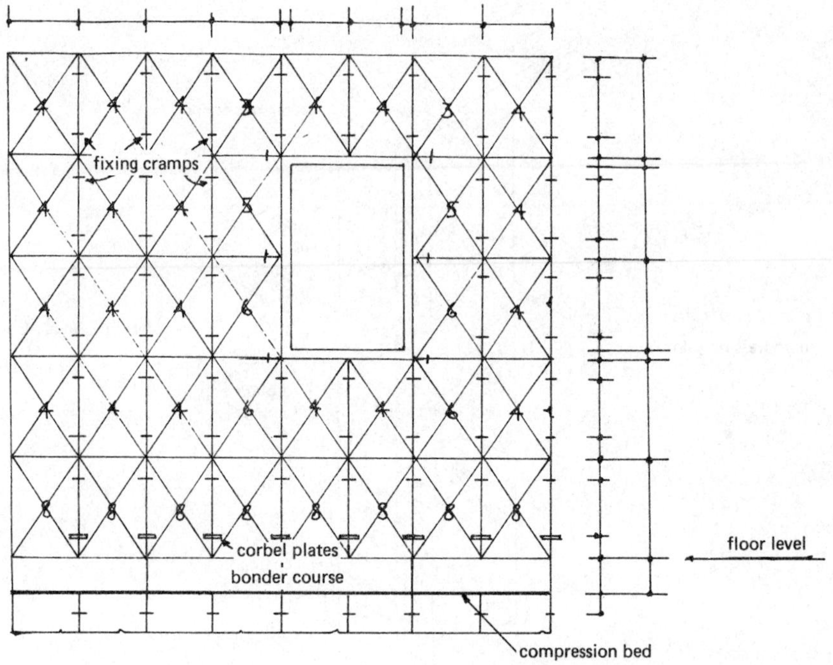

Figure 1.14 Elevation of grid for fixings

general rule for sizes of the above materials is that the ratio of length to breadth should not normally exceed 3:1, and the smallest face dimension should not exceed 600 mm.

Concrete Slabs

When formed with an aggregate size of 19 mm the minimum thickness may be reduced to 50 mm. The

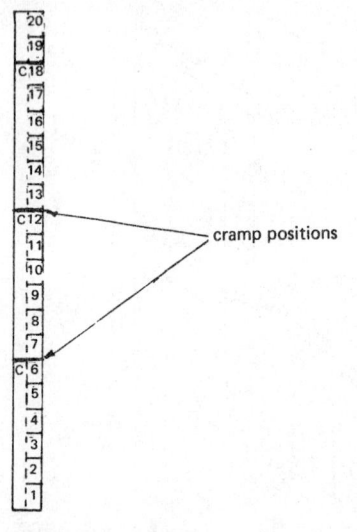

Figure 1.15 Gauge staff for cladding and brickwork

dimensions of concrete units vary according to the size of aggregate, thickness of slab and the amount of reinforcement within the unit.

FAILURES IN CLADDING

There are three main failures in cladding, which are attributed to a variety of causes, and which should be considered both at the design stage and during fixing operations. The failures are as follows.

(1) Spalling at horizontal joints, which occurs in large areas of cladding and cladding between mullions. The failure may be due to deformation of the structure or to differential thermal movement.
(2) Spalling at vertical joints, which often occurs when there are long uninterrupted lengths of cladding. The cause is termed 'creep', which is a progressive distortion induced by sustained stress or differential thermal movement.
(3) Displacement, which may be caused by failure of the fixings, deformation of the structure or because slabs were fixed directly on to a concrete backing.

Use of Bonder Courses

It is a general recommendation that the weight of each storey height of external cladding is transferred

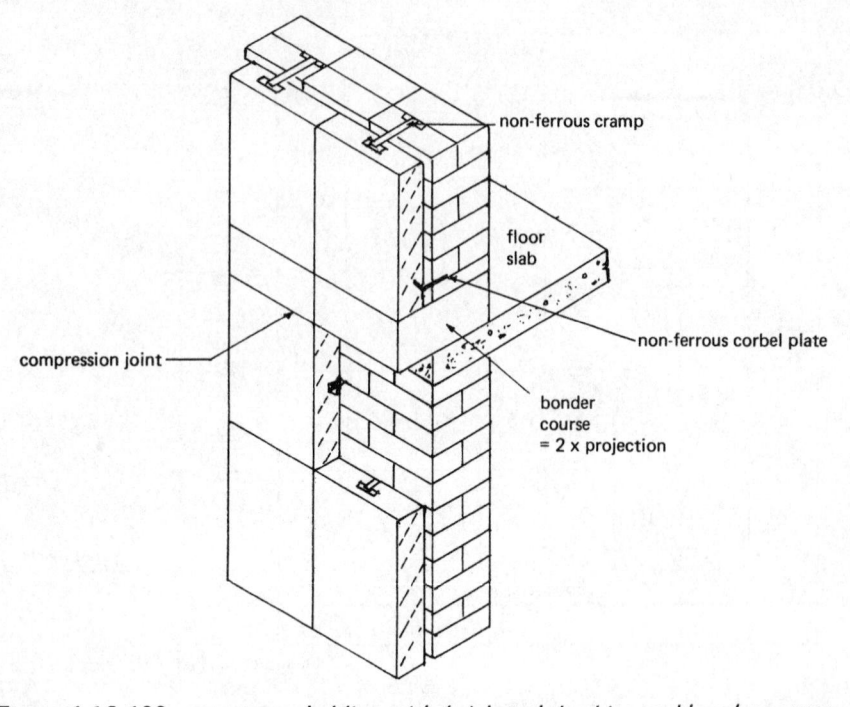

Figure 1.16 100 mm stone cladding with brickwork backing and bonder course

on to the structural framework (figures 1.16 and 1.17). The bonder course should have sufficient thickness to bear on the floor slab or beam, or it can be built into the backing wall. The bearing of a bonder course on the structural frame must be at least twice the pro-jection and not less than 112 mm (figures 1.18 and 1.19).

When cladding slabs are 100 mm thick a concrete boot lintel can be used. This is cast *in situ* with the concrete frame and occurs at each floor level. Concrete

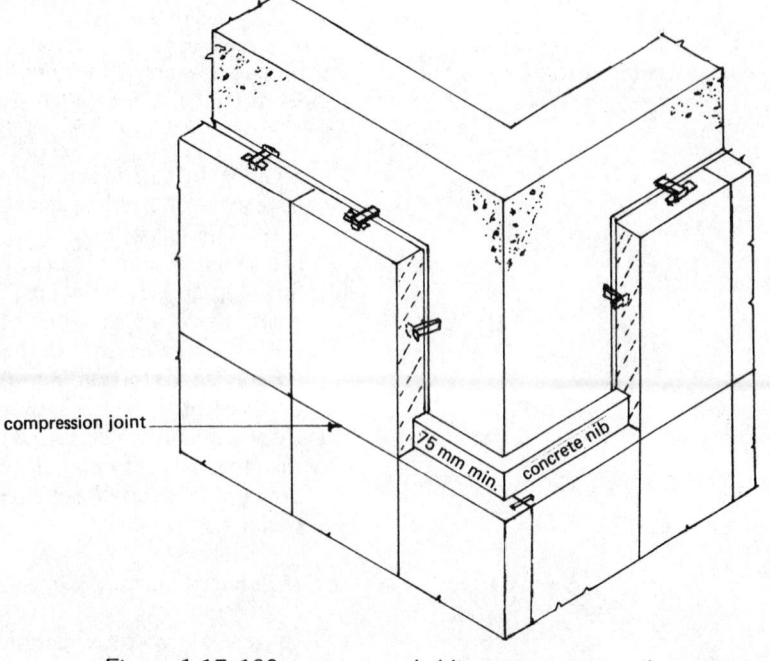

Figure 1.17 100 mm stone cladding on concrete nib

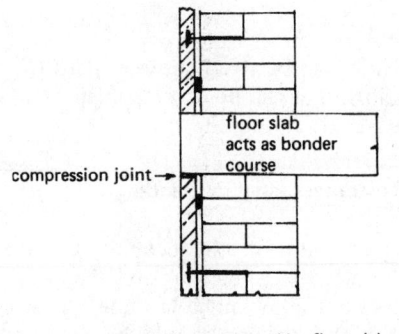

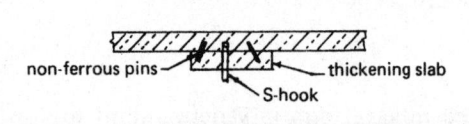

non-ferrous pins — thickening slab
S-hook

method of increasing thickness for thin slabs

compression joint →

floor slab acts as bonder course

thin cladding supported by floor slab

Figure 1.18

nibs can also be used. The cladding slabs are rebated to rest on the nib, which in turn transfers the weight to the structure. Nibs should be at least 100 mm in depth and should project at least 62 mm from the concrete face (figure 1.17).

An illustration of stone cladding at a return quoin is given in figure 1.20.

MOVEMENT JOINTS

Compression Joints

It is essential to provide a compression joint at every storey height; this will minimise the risk of spalling and displacement of the cladding slabs. The compression joint takes up the vertical movement of the building frame and relieves the pressure on the cladding (figures 1.16 and 1.19).

The position for a compression joint is immediately below the bonder course or concrete nib. It may also be incorporated at any position required by the structural engineer.

Expansion Joints

These are vertical joints placed at specified positions in the cladding, and their purpose is to allow for lateral movement. It is not possible to provide a rigid formula for the location of this type of joint; the number of joints and positions are determined by

(1) the material used for the structure
(2) the area of cladding involved
(3) the aspect of the walling
(4) end restraint on the cladding.

Materials Used to Form Compression and Expansion Joints

The materials used for both the above joints should be capable of accepting movement and must be of

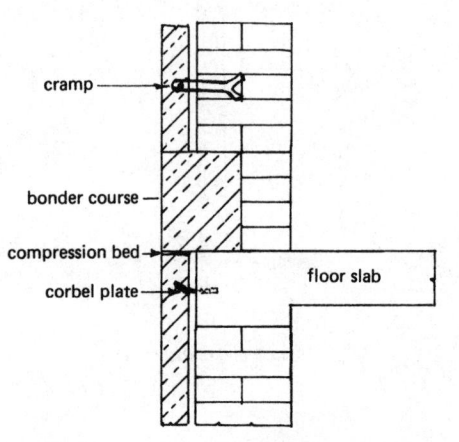

cramp —

bonder course —

compression bed →

corbel plate →

floor slab

Figure 1.19 65 mm stone cladding supported by bonder wrapover course

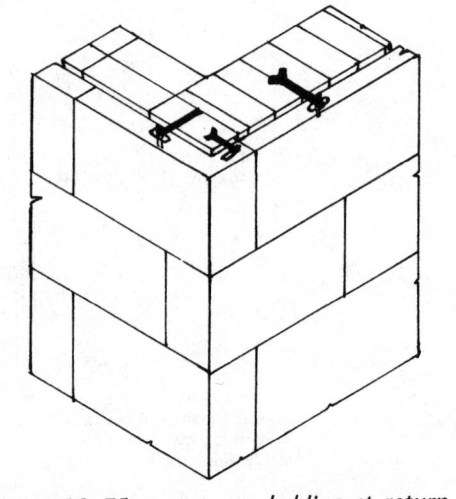

Figure 1.20 75 mm stone cladding at return quoin

non-hardening compounds, which may incorporate a sealer to resist moisture penetration.

Butyl Rubber

This is obtained in tape form in various widths to suit the thickness of the slabs used. The tape is placed on the dry, clean bedding surface of the slabs, and it must be ensured that it is kept back from the surface of the cladding units by at least 12 mm. When used for an expansion joint, it must be applied before the slab is fixed on the mortar bed.

Polysulphide Compounds

The compound is 'gunned in' under pressure after the slabs have been fixed. The joint is left open to allow for the compound and the slabs are supported on asbestos or hessian pads or plugs. Great care is required when applying the compound, otherwise face-staining of the cladding can occur.

Bituminised Foamed Polyurethane

Before applying the jointing material an adhesive must be painted on to the bedding surface of the cladding slabs. The use of this type of joint often causes delay. The foamed polyurethane cannot be placed on the adhesive until a period of at least 45 minutes has elapsed. The tape or strip should be kept back from the cladding face by at least 12 mm.

Copper Springs

These can be fixed between slabs to form expansion joints, and can be accompanied by any of the above materials.

Tolerances Used in Cladding

The amount of tolerance normally specified between the face of the backing and the back of the cladding does not allow much latitude, therefore it is essential that the backing should be constructed accurately with tolerances not exceeding ±6 mm in 5.0 m for both vertical and horizontal alignment.

The general accepted tolerance for stone cladding is

Length	Tolerance
Up to 3.0 m	±6 mm
3.0 to 5.0 m	±9 mm
5.0 to 7.0 m	±12 mm

Copings and Soffits

Copings formed from sandstone and limestone require cramps to be fixed into the top of the coping slabs and should be provided with a damp-proof course (d.p.c.). Slate copings do not require a d.p.c. and, when placed on top of a concrete or brick-backed wall, they are secured by cramps or bolts fixed into

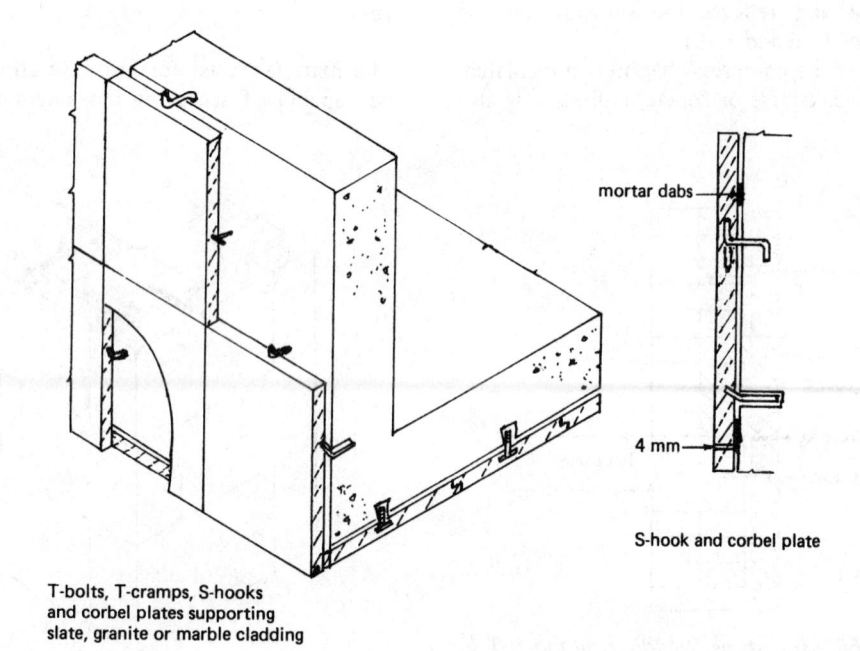

T-bolts, T-cramps, S-hooks
and corbel plates supporting
slate, granite or marble cladding

mortar dabs

4 mm

S-hook and corbel plate

Figure 1.21

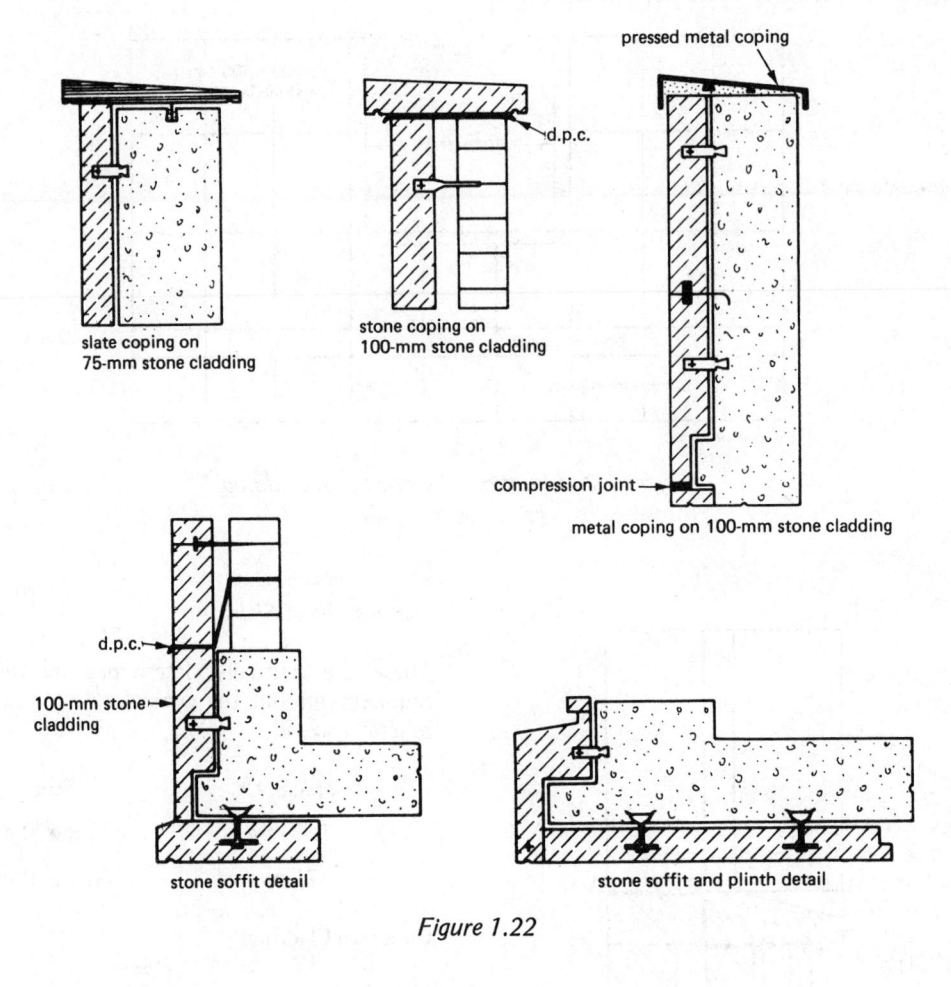

slate coping on
75-mm stone cladding

stone coping on
100-mm stone cladding

d.p.c.

pressed metal coping

compression joint

metal coping on 100-mm stone cladding

d.p.c.

100-mm stone
cladding

stone soffit detail

stone soffit and plinth detail

Figure 1.22

the top of the backing wall and into slots formed on the underside of the coping. Metal copings are fixed by patent clips; lugs are welded to the underside of the coping and secured in the mortar bed. Soffits are secured by adjustable bolts, which are fixed into the concrete beams or lintels above (figures 1.21 and 1.22).

DETAILS OF STONE AND CONCRETE CLADDING

Stone-faced Cladding

Stone-faced cladding slabs are a relatively new development in cladding. The advantages of this form of construction are

(1) there is a reduction in the quantity of stone required
(2) lightweight aggregate can be used as the backing for the stone facing slabs

(3) insulation can be increased, with a reduced wall thickness
(4) there is a reduction in labour costs

Requirements for Stone-faced Cladding

It is essential that the minimum thickness of the stone facing should be not less than 25 mm and the maximum area of a stone facing should not exceed 0.47 m² (figures 1.23–1.25).

The type of mortar recommended for use with this type of cladding is seven parts stone dust, four to five parts hydrated lime and one part white cement, but this ratio can vary according to the type of stone used. Both vertical and horizontal joints should not exceed 6 mm.

Fixing is achieved by non-ferrous channels and anchors formed in the concrete backing of the slabs, which is then secured in the backing wall. A 12 mm gap should be provided, which is later filled with cement grout.

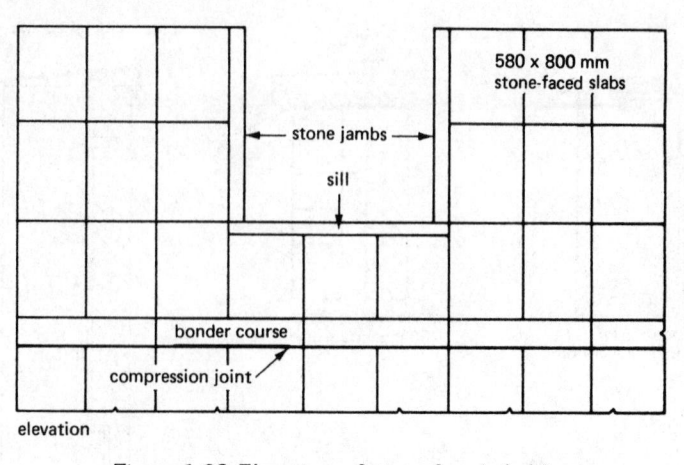

Figure 1.23 Elevation of stone-faced cladding

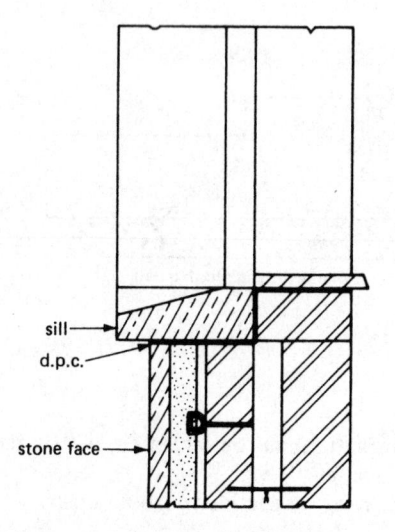

Figure 1.24 Section through sill

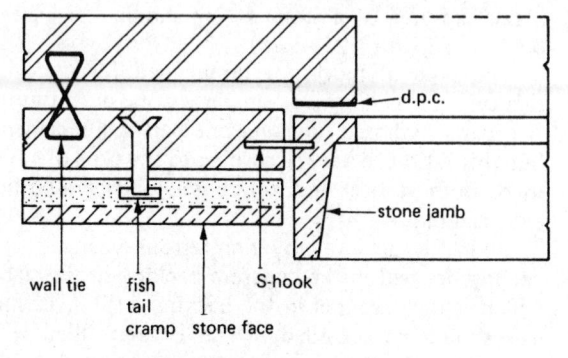

Figure 1.25 Jamb detail

Dimensions of Slabs

These are normally determined by the designer's requirements but the general rule for the maximum practical size is

Thickness	Size
150 mm	1650 x 600 mm
75 mm	1200 x 600 mm

Concrete Cladding

Manufacture

Precast concrete slabs for use as cladding are formed with two mixes in each mould. The mix used for forming the face of the slab is determined by

(1) the face appearance required by the designer
(2) the amount of weathering and degree of exposure.

For normal requirements the ratio of cement to fine aggregate should be not less than 1:3 and not greater than 1.1½.

Slab Dimensions

When slabs are formed with 19 mm aggregate and are of constant thickness, the minimum recommended thickness of slab is 50 mm; the maximum area for this thickness is 0.84 m² and the maximum dimension of the slab should not exceed 1.05 m.

When the size of aggregate exceeds 19 mm the thickness of the slab must also be increased.

It is recommended that when the weight of the slab exceeds 55 kg fixing devices should be inserted into the slabs and mechanical lifting equipment should be used.

Patent plastic inserts are also used in the concrete backing to provide holes for dowels at the position of window openings.

Considerations Affecting Tolerances

Before tolerances can be determined for concrete slabs the following factors must be considered

(1) inaccuracy at the mould assembly stage
(2) the production of slabs which are not square
(3) the sides of slabs may be twisted or warped
(4) badly shaped moulds which produce concave or convex cross-sections in the slabs

Manufacturers of concrete cladding slabs usually inform the designers of the tolerances that are required. Normally the general guide for tolerances of concrete slabs is as in table 1.1.

Table 1.1

Length	Tolerance	Bow	Twist
Up to 3.0 m	±6 mm	5 mm	5 mm
3.0 to 5.0 m	±9 mm	6 mm	6 mm
5.0 to 7.0 m	±12 mm	6 mm	6 mm

Slab size is determined by first obtaining the joint thickness required.

Example 1.1

Slab size is 2.4 m long and the thickness of joint required is 10 mm. The maximum size for the finished slab is obtained by subtracting the joint thickness from the slab length.

$$2400 - 10 \text{ mm} = 2390 \text{ mm}$$

Minus the tolerance in table 1.1 = 6 mm

Therefore

size of the slab when produced = 2384 mm

Reinforcement

This is always placed in the backing concrete of the slab. It should have 50 mm of surface cover and 50 mm of cover at the sides.

Reinforcement is never placed between the two layers of concrete. The dimensions of the mesh are determined by the thickness of concrete.

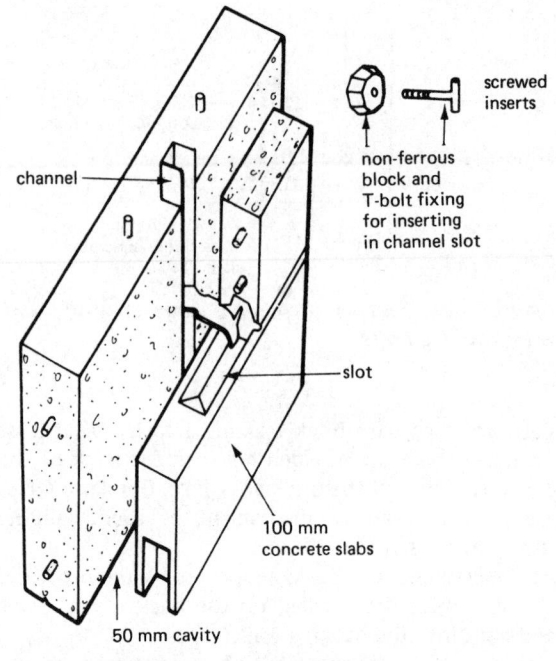

Figure 1.26 Concrete cladding secured by metal channel and one-piece cramp

Slots for fixing are always fixed into the concrete backing; polystyrene blocks or inserts are also placed in the backing, and can be taken out later to accommodate metal fixings. With large concrete units it is also the practice to insert adjustable levelling bolts and screwed metal inserts which can be used for lifting purposes (figures 1.26–1.28).

Fixing Techniques

The method of fixing used for concrete slabs is the same as for stone slabs, that is, non-ferrous metal ties, anchors and cramps are used. When concrete

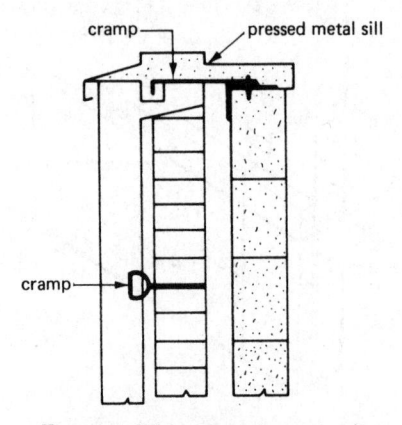

Figure 1.27 Section through sill

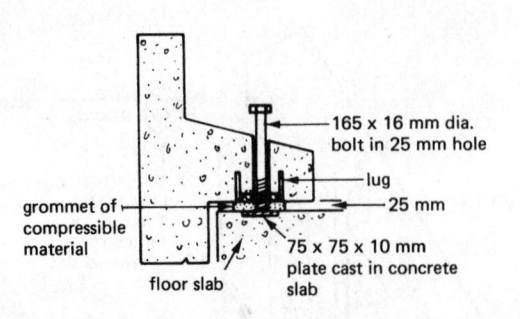

Figure 1.28 Section through a large concrete slab with levelling bolts

slabs are used with brick backing a cavity of at least 18 mm should be provided between the back of the concrete slab and the brickwork face. Bonder courses are used to transfer the weight of each cladding storey to the structural frame.

Concrete slabs are also supported by the provision of nibs; these are formed on the back of the slabs and built into the backing wall.

When the backing wall is of cast-*in-situ* concrete, slots are formed in the backing. The slots are temporarily filled with polystyrene and metal slots are inserted later; these then accommodate cramps which are secured into mortice holes in the back of the concrete slabs. Channels fixed into the concrete backing are also a common method of securing cramps for the concrete slabs (figure 1.29).

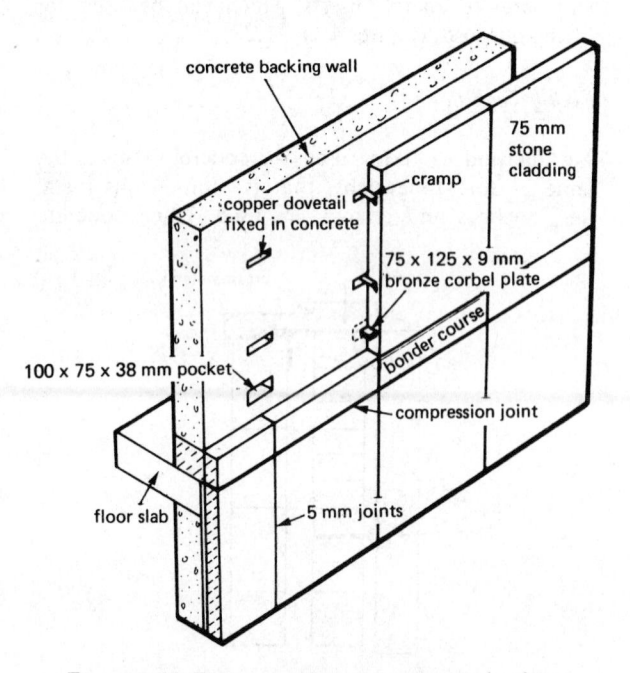

Figure 1.29 Fixing positions in concrete backing

TREATMENT OF JOINTS

There are four types of vertical joint that may be formed between concrete cladding slabs; these joints may also be used on other forms of cladding where desired by the designer. The joints are

(1) gap-filled joints
(2) sealed joints using cover strips
(3) lapped joints
(4) open-drained joints.

Gap-filled Joints

The jointing material must be capable of providing adhesion with the cladding slabs and should not contract or suffer fracture. With this type of joint it is necessary to provide a sealant of butyl mastic or gunned-in polysulphide compound (figure 1.30).

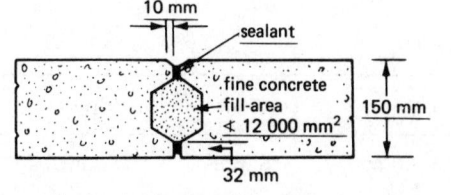

Figure 1.30 Plan of gap-filled joint

Sealed Cover Strip Joints

This type of joint is formed by inserting a clip spring between the slabs and attaching a metal cover strip to the spring. It is essential for the cover strip to lap over the slabs by at least 12 mm on each side. A mastic seal is formed on the inside of the joint to prevent any moisture penetration (figure 1.31).

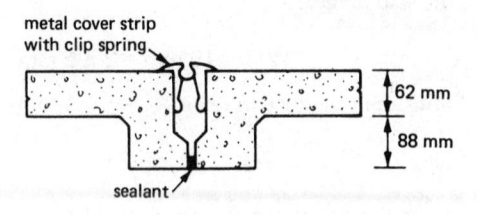

Figure 1.31 Plan of cover strip joint

Lapped Joints

Lapped joints are obtained by forming a rebate on the sides of the slabs. Before fixing the slab, a compressible strip of butyl rubber is placed between **the** rebates of the slabs. This method of jointing **requires**

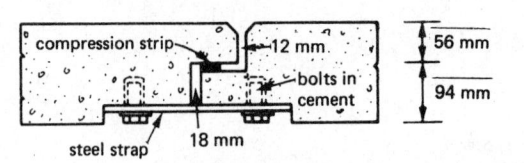

Figure 1.32 Plan of lapped joint

considerable care, otherwise the joint strip may be damaged or displaced during fixing (figure 1.32).

Open-drained Joints

To form open-drained joints it is essential to provide a width of joint of at least 9 mm, otherwise blockages may occur. Grooves should be formed in the side of the slabs and a baffle strip should be inserted into the second groove. A sealant can be used as added prevention at the back of the joint (figure 1.33).

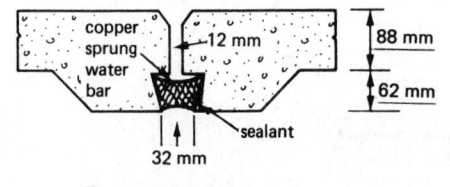

Figure 1.33 Plan of open joint

Cavities are normally left unfilled; this trend is now becoming common practice. When cavities are filled with cement grout any voids that may occur allow moisture to accumulate and penetrate the backing walls.

Horizontal Joints

These are formed with mortar and compressible strips or mastic. Lightweight slabs are often provided with rebated joints, formed with cement mortar and strips. It is obviously necessary to provide a weak mortar with a minimum water content (figure 1.34).

Large, heavy slabs are often provided with a rebated joint having an upstand of at least 75 mm. This is left open to prevent any capillarity (figure 1.35).

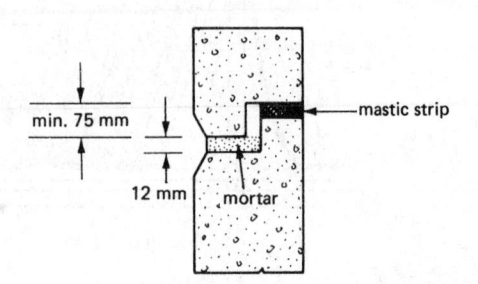

Figure 1.35 Section of horizontal joint for thick slabs

PROTECTION OF CLADDING

Slabs

When slabs are to be stacked on working platforms it is essential to increase the strength of the scaffold by the addition of extra ledgers, putlogs and, if necessary, standards.

The slabs should always be stacked on compressible layers of felt or soft boardings. Softwood compressible spacers should be placed between each slab and the next and a clean, lightweight sheet should be used to cover the slabs and ensure complete dryness before fixing (figure 1.36).

When work stops, either at break or at the end of the day, it is very important to provide complete protection for all cladding and especially for newly fixed slabs. This can be done with temporary timber frames and clean, lightweight sheeting, which should be wrapped over the top of the walling at least 300 mm and taken down the face of the cladding to platform level; adequate clearance of the face and security against displacement must be ensured (figure 1.37).

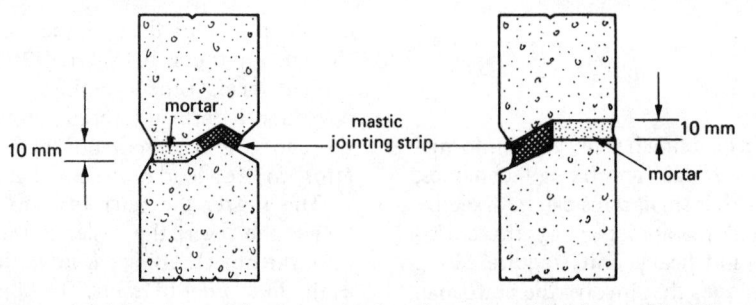

Figure 1.34 Sections of horizontal joints for slabs of thickness 62–75 mm

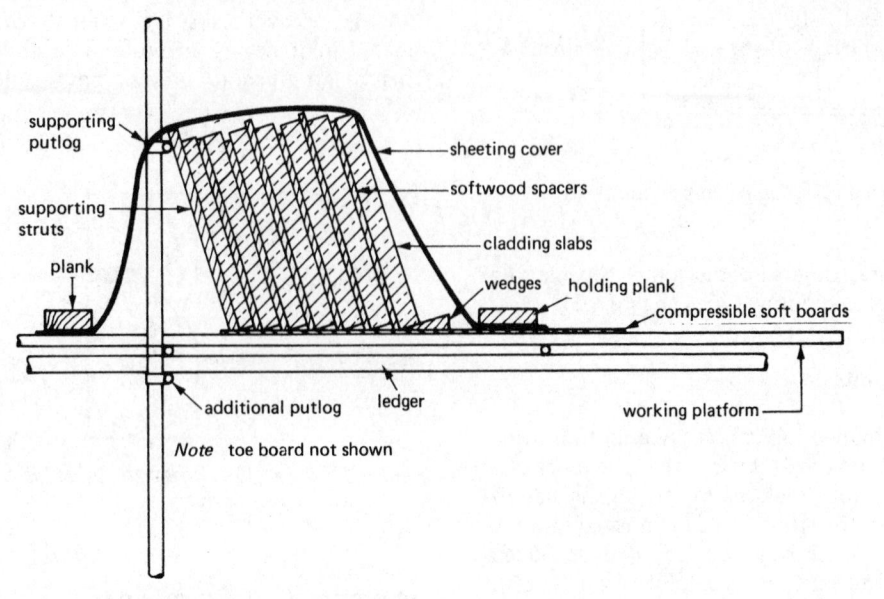

Figure 1.36 Method of stacking and protecting cladding slabs on working platforms

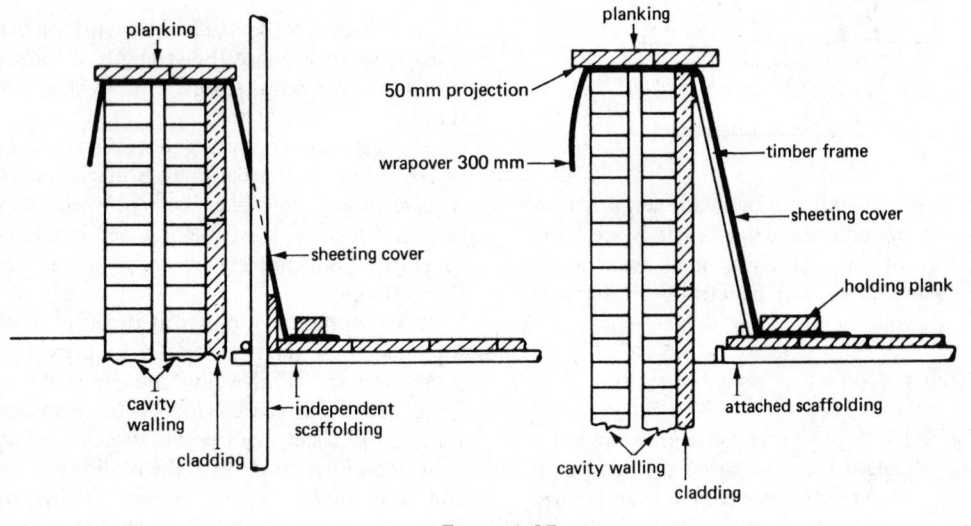

Figure 1.37

MECHANICS

The Lever

The present chapter on cladding has provided information on the dimensions of slabs and fixing techniques. It will be appreciated that small slabs can be fixed by manual operations and manipulation by the craftsman, but large slabs and heavy units require easing and positioning techniques that involve the craftsman in using a pinchbar or crowbar, usually covered with a 'sock' placed on the end, that is, under the centre of the end of the slab (figure 1.38). Some bars have a formed heel, others require a pivot block. This operation is termed levering, and we should consider the bar as a machine requiring a downward force or effort to lift a load or weight (figure 1.39).

The nearer the fulcrum is to the object and the longer the lever, the easier is the effort required to raise the object. Using a lever in this manner is known as the first order of levers. The object is raised because the 'clockwise moment about the fulcrum is greater

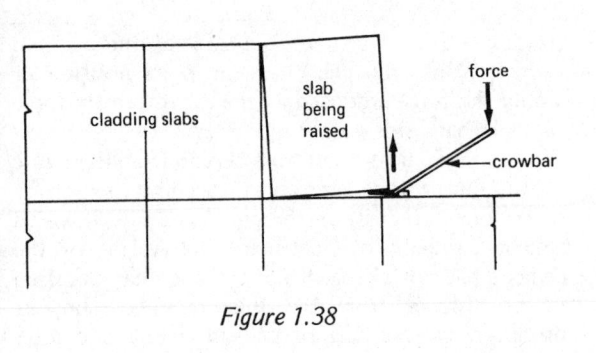

Figure 1.38

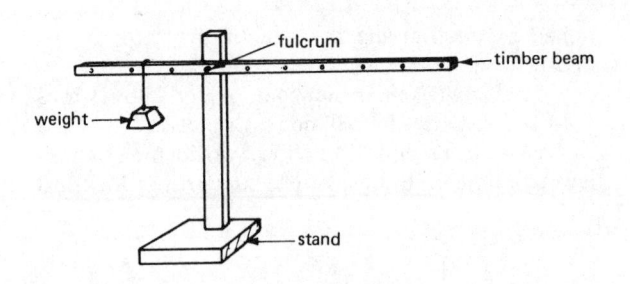

Figure 1.40

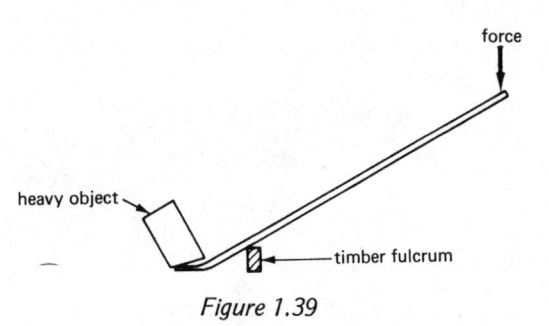

Figure 1.39

than the anticlockwise moment'. Figure 1.40 shows the apparatus for a simple experiment to study the first order of levers.

Experiment

Requirements for the experiment are a timber beam, measuring 38 x 25 mm, 1 m in length, with holes every 100 mm; one winged nut and bolt, a stand and various weights.

Method Set up the apparatus as shown in figure 1.40, fixing the fulcrum in a hole between positions 3 and 8 inclusive. Place a small weight on the short length of the beam to balance the self weight of the beam.

Start by placing a force of 40 N (4 kg) at 600 mm to the right of the fulcrum and a force of 80 N (8 kg) at 300 mm to the left of the fulcrum. The beam will be seen to be in balance and this is due to the fact that the clockwise moments (CM) equal the anticlockwise moments (ACM).
That is

$$40 \text{ N} \times 600 \text{ mm} = 24000 \text{ N mm (CM)}$$
and
$$80 \text{ N} \times 300 \text{ mm} = 24000 \text{ N mm (ACM)}$$

To continue, place a force of 30 N (3 kg) at 500 mm to the right of the fulcrum. A force of 50 N (5 kg) is then placed at 300 mm to the left to provide the required balance, that is, the CM must equal the ACM

That is

$$30 \text{ N} \times 500 \text{ mm} = 15000 \text{ N mm (CM)}$$
and
$$50 \text{ N} \times 300 \text{ mm} = 15000 \text{ N mm (ACM)}$$

A table can now be drawn up as shown in table 1.2 and different forces applied at varying positions on the beam.

Thus it can be stated that a moment is the turning effect of a force about a point and its value is deter-

Table 1.2

Left-hand side, ACM				Right-hand side, CM			
Mass (kg)	Force (N) (mass x g)	Distance from fulcrum (mm)	Moment (N/mm)	Mass (kg)	Force (N) (mass x g)	Distance from fulcrum (mm)	Moment (N/mm)
4	40	600	24000	8	80	300	24000
3	30	500	15000	5	50	300	15000

It is assumed that g is 10 m/s^2.

mined by multiplying the size of the force by its distance from the point.

Forces are given in newtons (N) or kilonewtons (kN) and distances in millimetres or metres.

Figures 1.41 and 1.42 show two identical cantilevered timber beams, each supporting identical

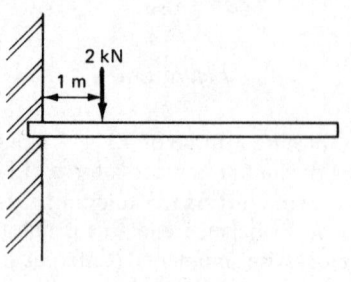

Figure 1.41

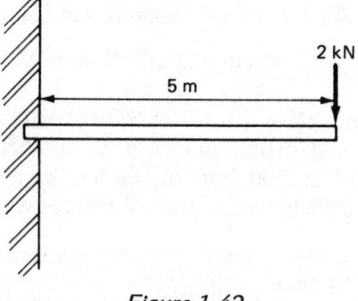

Figure 1.42

forces of 2 kN. It is obvious that if either beam were to fail under load, it would be the one in figure 1.42, since the force is acting at a greater distance from the beam support. The moment produced in figure 1.41 is

force x distance = 2 kN x 1 m

= 2 kN m

The moment produced in figure 1.42 is

force x distance = 2 kN x 5 m

= 10 kN m

Figure 1.43 shows a heavy plank supported by a

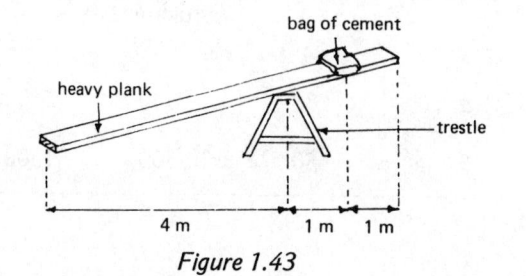

Figure 1.43

trestle. A bag of cement is placed at the position shown. Would the plank remain in its position or would the force produced by the bag of cement force the right-hand side down?

This would be difficult to ascertain from the figure shown, but if the cement were moved to the extremity of the plank, that is, on the right-hand side, it would almost certainly over-balance. The weight of the cement has not changed but a change has occurred in its distance from the fulcrum. The moments produced by the bag of cement before and after moving are

position 1

moment = force x distance

= 500 N x 1 m

= 500 N m

position 2

moment = force x distance

= 500 N x 2 m

= 1000 N m

(assuming g = 10 m/s^2)

Figure 1.44 represents a beam resting on a fulcrum and carrying different forces at different positions from the fulcrum. Ignoring the self weight of the beam, will this arrangement produce a balance or will one end be lowered to the ground?

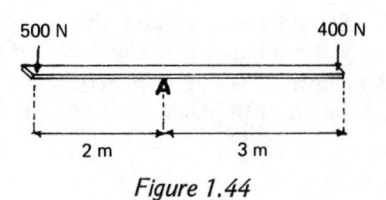

Figure 1.44

This can be determined by experiment or calculation. The calculation method would require the calculation of moments produced on each side of the fulcrum. Figures 1.45 and 1.46 show diagrams of the anticlockwise and clockwise moments respectively.

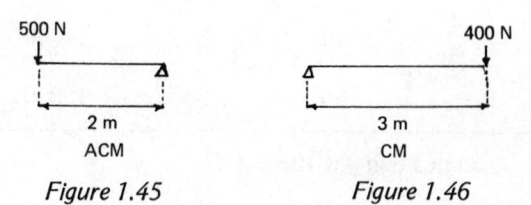

Figure 1.45   *Figure 1.46*

Anticlockwise moment = force x distance

$$= 500 \times 2$$

$$= 1000 \text{ N m}$$

Clockwise moment = force x distance

$$= 400 \times 3$$

$$= 1200 \text{ N m}$$

This result would indicate that the right-hand end of the beam would be lowered and resting on the ground.

First Order of Levers

Example 1.2

Calculate the effort required to raise the load shown in figure 1.47.

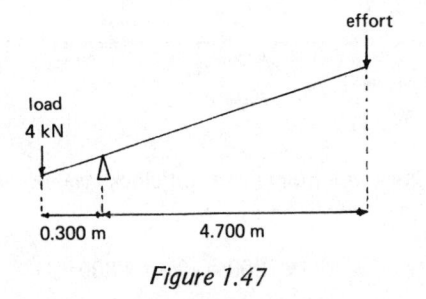

Figure 1.47

Clockwise moments (CM) = anticlockwise moments
(ACM)

Therefore

$$\text{effort} \times 4.7 = 4 \times 0.3$$

$$= 1.2$$

Therefore

effort $= \dfrac{1.2}{4.7}$ (divide both sides by the the coefficient of 'effort' and cancel)

$$= 0.26 \text{ kN}$$

Thus an effort of 0.26 kN (260 N) will balance the load and any addition to this will cause the load to rise.

Example 1.3

Calculate the effort required to raise the load shown in figure 1.48

$$\text{CM} = \text{ACM}$$

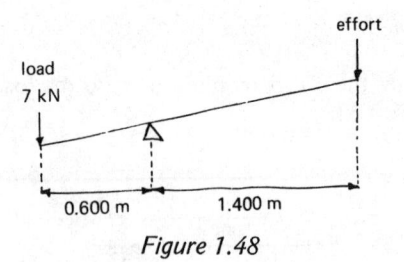

Figure 1.48

Therefore

$$\text{effort} \times 1.4 = 7 \times 0.6$$

Therefore

$$\text{effort} \times 1.4 = 4.2$$

Therefore

$$\text{effort} = \frac{4.2}{1.4}$$

$$= 3 \text{ kN}$$

Therefore any effort over 3 kN will cause the load to rise.

Second Order of Levers

A method of raising a heavy object that is already clear of the ground is to place a plank or crowbar under the object with one end resting on a firm surface and lift it upwards to form contact with the object. An upward force is then applied at the other end of the lever (figure 1.49).

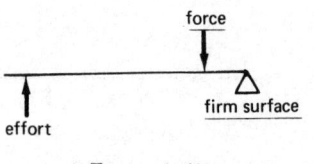

Figure 1.49

The wheelbarrow is an excellent example of the use of the second order of levers (figure 1.50).

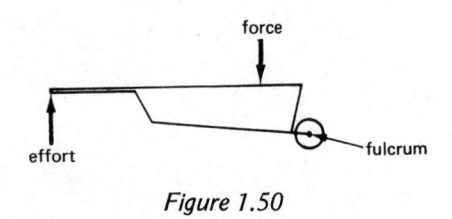

Figure 1.50

Example 1.4

Calculate the effort required to raise the load shown in figure 1.51.

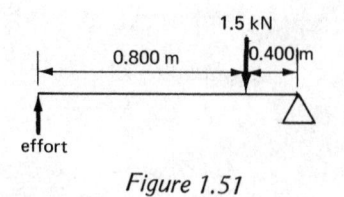

Figure 1.51

Note   Moments are always calculated from the fulcrum.

Clockwise moments = anticlockwise moments

Therefore

$$\text{effort} \times 1.2 = 1.5 \times 0.4$$
$$= 0.6$$

Therefore

$$\text{effort} = \frac{0.6}{1.2}$$

$$= 0.5 \text{ kN}$$

Any effort greater than 0.5 kN will raise the load.

Third Order of Levers

Figure 1.52 shows the arrangement for the third order of levers. In this example the effort required to raise the load is always greater than the load itself.

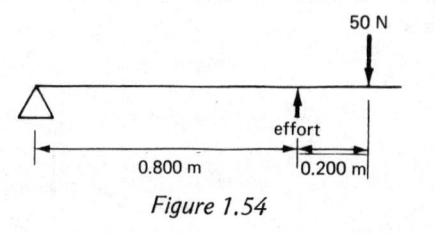

Figure 1.52

Figure 1.53 shows this to be a very convenient arrangement.

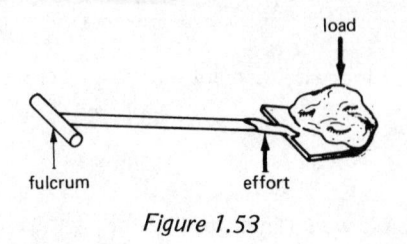

Figure 1.53

Example 1.5

Calculate the effort to balance the load shown in figure 1.54.

Figure 1.54

Clockwise moments = anticlockwise moments

Therefore

$$50 \times 1000 = \text{effort} \times 800$$

Therefore

$$50\,000 = \text{effort} \times 800$$

Therefore

$$\text{effort} = \frac{50\,000}{800}$$

$$= 62.5 \text{ N}$$

It will be noted that the effort is in excess of the load.

2
DECORATIVE AND FUNCTIONAL FEATURES

In the construction industry designers are now taking advantage of the modern technology that enables the manufacturers to supply materials considerably superior in quality to those available years ago. The result is that many newly constructed buildings have façades consisting of contrasting-coloured bricks with surface textures chosen to suit the particular environment.

While colour and texture have improved we find that many buildings lack the decorative aesthetic qualities that are to be seen in older buildings; decorative face bonds are seldom used, and buildings now constructed with drab and plain face areas could be greatly enhanced by including decorative features in the brickwork.

STRING, DENTIL AND DOG-TOOTH COURSES

String Courses

String courses are used to provide distinct breaks in the façade of buildings. They form sub-divisions and interrupt the continuity of the face-work. A string course is a projecting course of decorative or contrasting brickwork, which may be in a different bonding arrangement from that used for the brickwork face. Care should be taken to ensure that there is a minimum amount of lap between the projecting or oversailing courses and the brickwork above and below; the lap should be not less than 28 mm.

Dentil Courses

Dentil courses are courses of bricks formed with every alternate brick projecting beyond the face of the walling, or every alternate brick set back from the brickwork face (figure 2.1). Dentil courses are built with header courses or bricks on end, with a projection or recess not greater than 28 mm, although 18 mm is normally used.

To increase the depth of the band or string course the dentil courses can be incorporated in the over-

sailing courses. This provides a very decorative band on the face of the walling and is often used externally at first-floor level. When building string courses, the eye-lines should be formed at the bottom arris of the first oversailing course and the top arris of the last course (figure 2.1).

Dog-tooth Courses

Dog-tooth courses are another means of forming a decorative effect on the face of brick walls; the dog-toothing is obtained by setting each brick at 45° to the wall face. Dog-tooth courses may be projecting, flush or recessed from the wall face. Bricks are used flat or on end (figure 2.2); they should be set to line and checked for accuracy and position with a triangular templet (figure 2.3).

To provide a more decorative effect and increase the depth of band, continuous courses of dog-toothing are often used (figures 2.2 and 2.4). It is also common practice to form dog-tooth courses between the oversailing courses. When bricks are laid flat to form dog-tooth courses in walls one brick thick, half bats are used to form the dog-toothing, thus allowing for a fair face to be obtained on the opposite face of the wall. When the wall is over one brick in thickness, a stretcher course can be used at the back of the dog-tooth courses.

DECORATIVE TREATMENT OF QUOINS

The most functional feature of any building is possibly the quoin, and yet in modern construction very little decorative treatment is given to this position. Creating a decorative effect with quoins, irrespective of the bonding arrangement of the walling, does not always increase the cost of labour. Drabness of the feature is completely erased and aesthetic qualities are provided that last for the life of the building.

There are two types of decorative quoin

(1) indented quoins
(2) rusticated quoins.

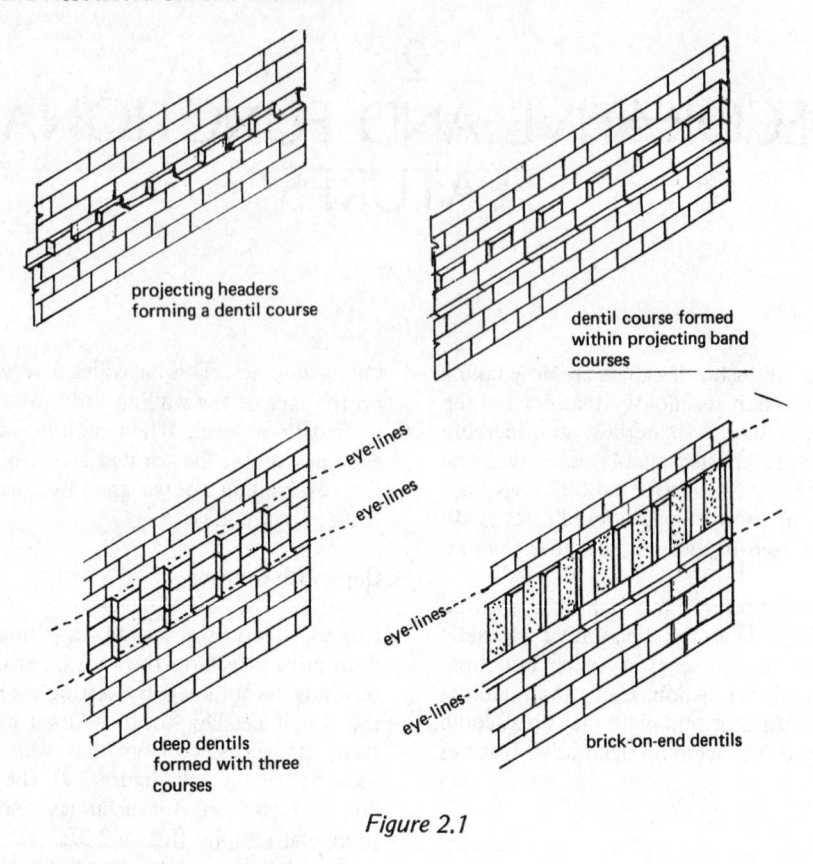

projecting headers
forming a dentil course

dentil course formed
within projecting band
courses

eye-lines

eye-lines

eye-lines

eye-lines

deep dentils
formed with three
courses

brick-on-end dentils

Figure 2.1

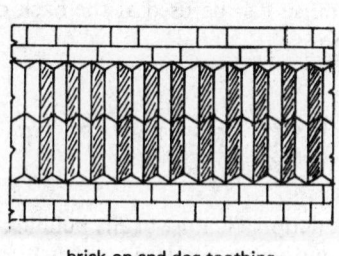

2-course dog-toothing

brick-on-end dog-toothing

Figure 2.2

Indented Quoins

Indented quoins are constructed with a recess in the
bricks forming the quoin. The total depth of the

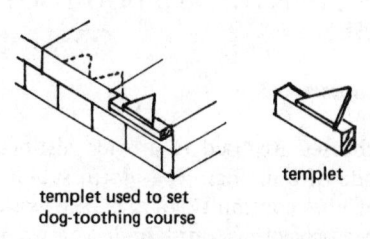

templet used to set
dog-toothing course

templet

Figure 2.3

indentation may be one or two courses. The depth of
the recess should not exceed 28 mm, although 19 mm
is the usual amount provided. This is obtained by
cutting only the quoin brick that forms the indentation (figures 2.5 and 2.6). The brick is reduced in
length and width by cutting off the amount of the
recess; remaining bricks forming the indentation
remain uncut. If the indentation is for one course, it
should occur every three courses, but when the
indentation is two courses deep, it should be formed
every four courses. This will provide the correct
bonding arrangement for the quoin whatever type of
bond is used for the walling (figure 2.7).

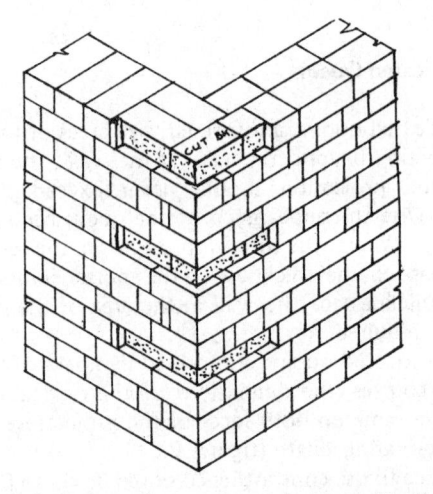

Figure 2.4

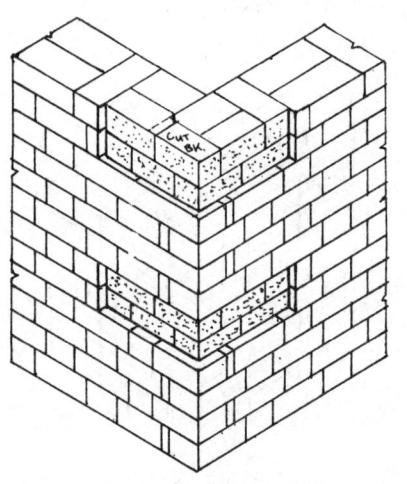

Figure 2.5 Indented or recessed quoin in Flemish bond

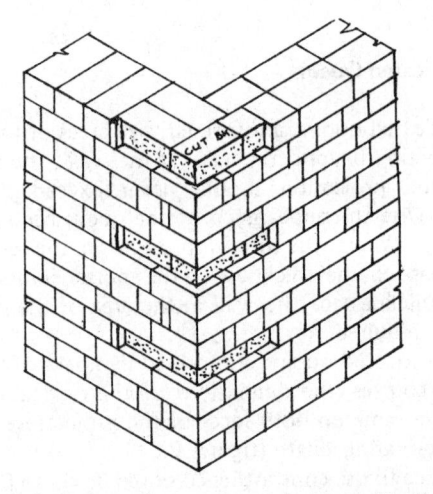

Figure 2.6 Indented or recessed quoin in English bond

Figure 2.7

Rusticated Quoins

Rusticated quoins are formed with areas of the quoin projecting beyond the face of the wall. The amount of the projection should never exceed 28 mm, otherwise the appearance of the quoin becomes too heavy. Where the bricks project on the quoin an increase in the thickness of the mortar joints within the thickness of the wall in the area of the quoin is often required.

Rusticated quoins are formed in blocks of three or four courses; the dimensions of the rustications can be the same on both faces or the projections can be of alternating length (figures 2.8–2.11). It is common practice to use contrasting-coloured bricks to form the rustications and indentations. This treatment increases the decorative features of the quoin.

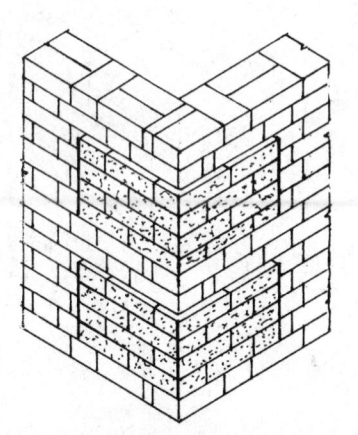

Figure 2.8 Rusticated quoin with equal projections on each face

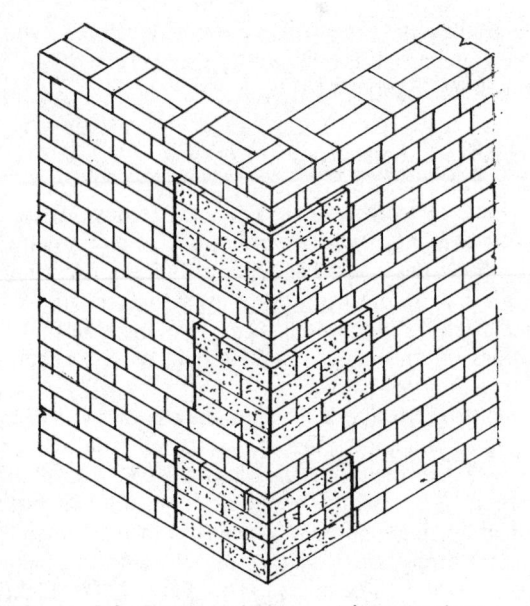

Figure 2.9 *Rusticated quoin with projections of alternating length*

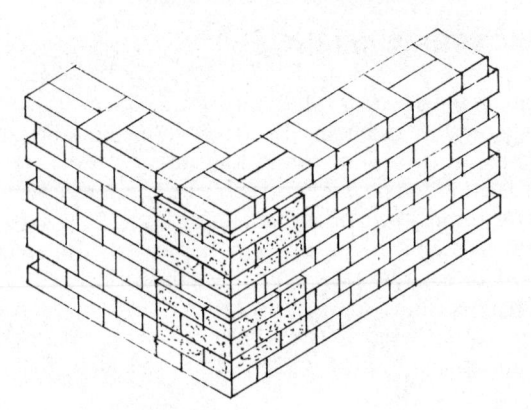

Figure 2.10 *Rusticated quoin in Flemish bond*

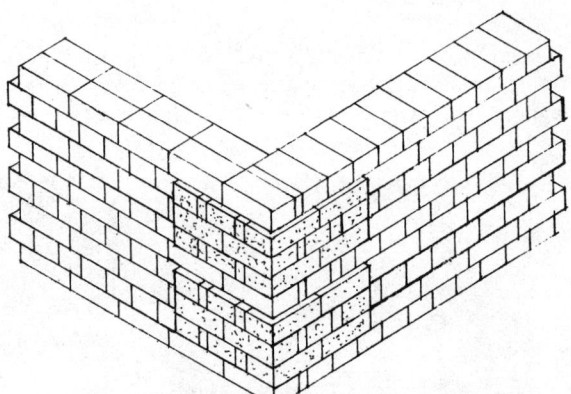

Figure 2.11 *Rusticated quoin in English bond*

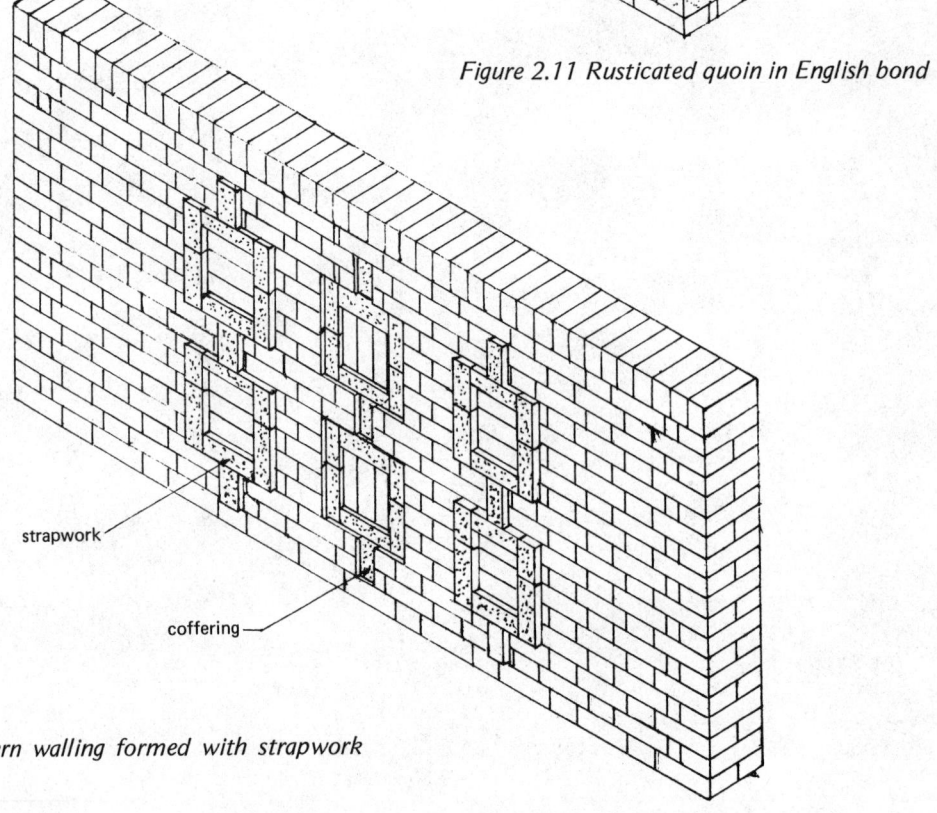

strapwork

coffering

Figure 2.12 *Pattern walling formed with strapwork and coffering*

PATTERN WALLING

On external walls the decorative effect can be enhanced by geometrical patterns formed on the face of the walling with contrasting-coloured bricks and projecting and recessed courses. When geometrical patterns formed with bricks projecting from the wall face are repeated either vertically or horizontally, the term used to describe the feature is strap work. Recessing fixed repeating patterns is termed coffering and it very often occurs when both of these decorative features are formed on the same wall face (figures 2.12 and 2.13).

Pierced Work

Pierced work is walling formed with repeating patterns, obtained by leaving voids or piercings in the walling. This type of wall is usually designed to form balustrading or screens. It is extremely effective and normally the wall is built with the same type of brick throughout (figures 2.14 and 2.15).

Decorative Internal Walls

Designers of modern brick buildings are now increasing the amount of exposed brickwork used internally, in both houses and other types of building. A study of internal decorative work shows that the use of geometrical patterns and light provides more aesthetic qualities for this type of walling than the use of contrasting-coloured bricks.

Figures 2.16 and 2.17 show the use of two types of zigzag bonding. The geometry of the design increases the depth of the decorative effect, while figure 2.18 shows the elevation of an internal wall built in herringbone bond which is formed to provide diagonal projections, even when built with the same type of brick throughout, as in figure 2.18. Light

Figure 2.13

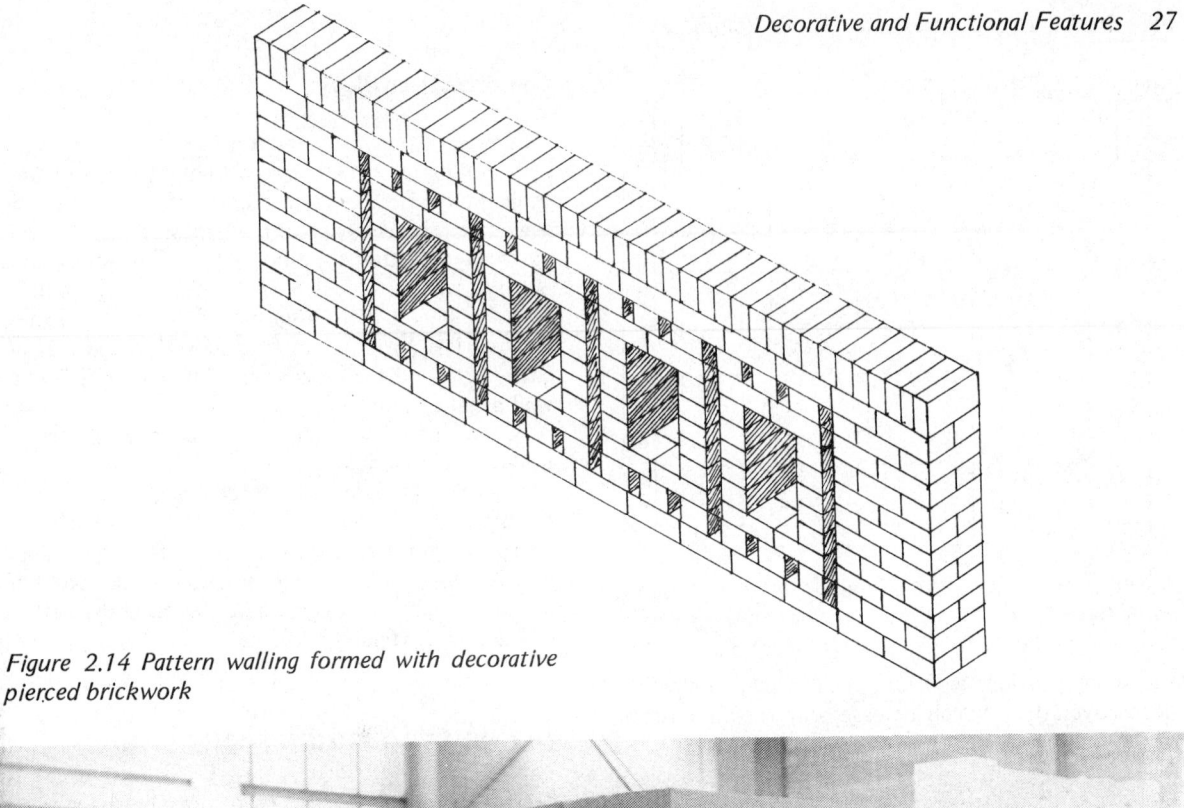

Figure 2.14 Pattern walling formed with decorative pierced brickwork

Figure 2.15

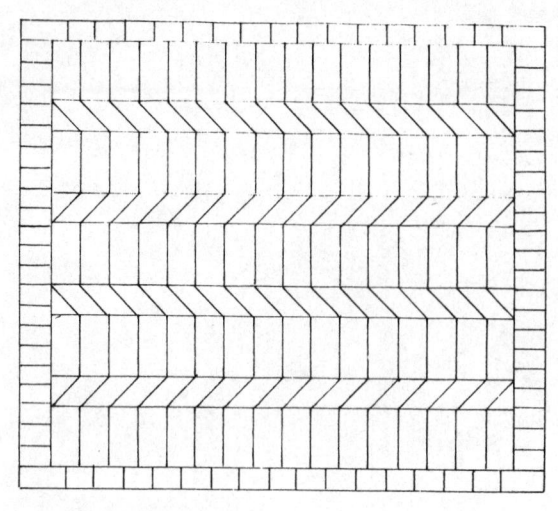

Figure 2.16 Single zigzag bond forming decorative panel (elevation)

and shade provide the walling with a very interesting decorative effect, which increases with the dimensions of the walling.

Lap Work

Lap work is another form of bonding arrangement used for decorative internal walls. Again the same type of brick can be used for the entire wall face, or bricks of similar colour can be used. The designer will again obtain the decorative effect with patterns of light and shade (figure 2.19).

Concrete Screen Blocks

This type of precast concrete block is now becoming very popular as a method of forming screen or balustrade walls. The concrete blocks are made with fine aggregate and cement, formed in steel moulds and compacted by the vibration method. The dimensions of the concrete blocks vary from 300 x 300 mm to 450 x 450 mm, with a normal thickness of 75 mm. Geometrical patterns are formed within the area of each block and the blocks can be used effectively for wall heights of up to 2.4 m.

When the blocks are used for long lengths of walling it is advisable to provide end and intermediate supports in the form of attached piers, which are also constructed of concrete blocks. These blocks are provided with a recess to accommodate the walling blocks. Plinths and copings can also be used to obtain increased stability and increase the decorative effect of the walling (figure 2.20).

DECORATIVE BRICK PANELS

The use of the panelled surface has long been a method of increasing the decorative qualities of walling. The use of the sunken or raised panel is often seen as a method of forming a feature in plain areas of walling and between piers. Placing of panels should be done with the utmost care. The dimensions and shape of the panel should coincide with the area of brickwork involved. Heavily sunken or too great projections

Figure 2.17 Double zigzag bond forming decorative panel (elevation)

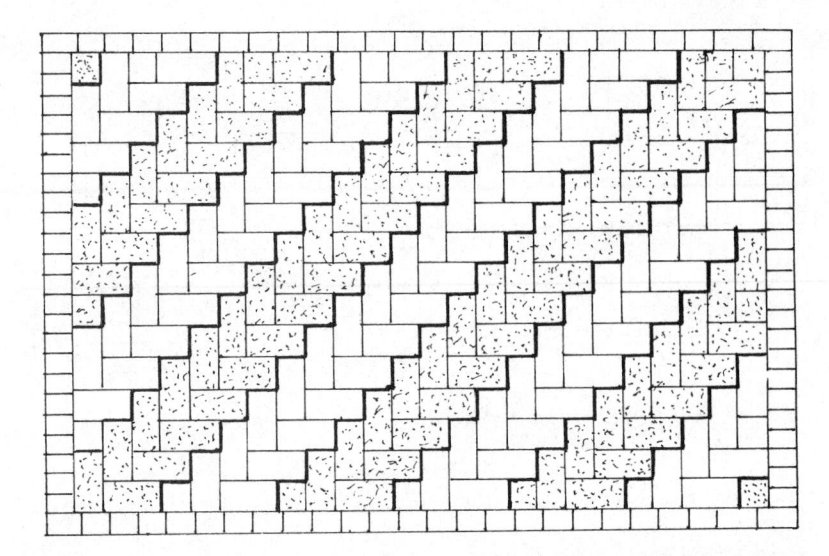

Figure 2.18 Internal decorative wall in herringbone bond with diagonal projections

Figure 2.19

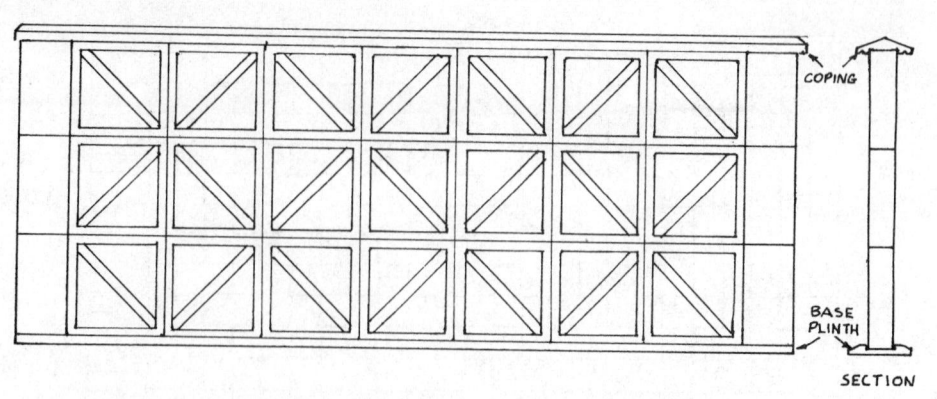

COPING

BASE
PLINTH

SECTION

Figure 2.20 Screen or balustrade block wall with end supports

often dilute the decorative effect intended for the feature.

A brick panel can be formed with a brick frame around the panel, or the panel can be formed within a recess in the walling. When a frame is used the base and sides should always be constructed before the panel is inserted, and should be formed with lines whenever possible (figure 2.21). If the panel is to project, the brickwork courses at the back of the recess are built using the bricks flat, but if a sunken or recessed panel is required, the courses of brickwork at the back of the recess should be formed with brick on

edge and block indenting into the brick walling on each side.

Bond for Panels

Brick panels can be formed with basket-weave or herringbone bonds. Although diagonal and other bonding arrangements can be used, the designer usually favours the former to provide the decorative effect required (figure 2.22).

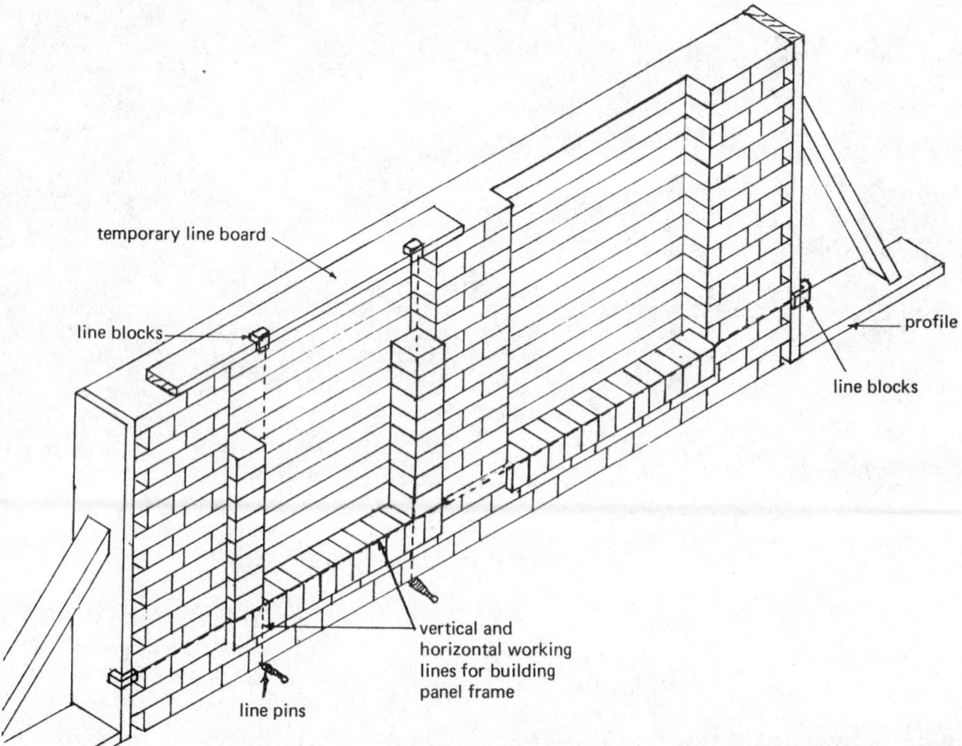

temporary line board

line blocks

profile

line blocks

vertical and
horizontal working
lines for building
panel frame

line pins

Figure 2.21 Method of building brick-on-edge frame around brick recess

Figure 2.22 Basket-weave panel with brick-on-edge frame

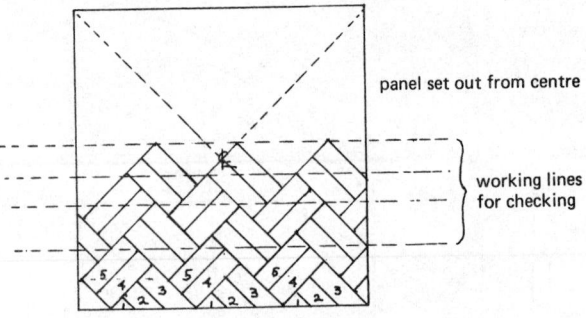

panel set out from centre

working lines for checking

Figure 2.24 Setting out a double herringbone square panel

Setting Out

Panels are usually set out according to

(1) the type of bond required
(2) the shape of the panel
(3) the dimensions of the panel.

When square panels are used, the setting out is always begun from the centre of the panel, whereas rectangular panels are always set out from the base line. This method is used to ensure that cuts, when required, are all the same and occur on both sides of the panel (figures 2.23—2.26).

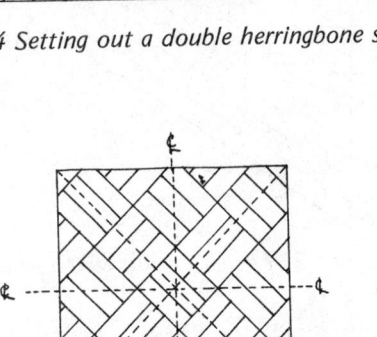

Figure 2.25 Setting out diagonal basket-weave in square panel

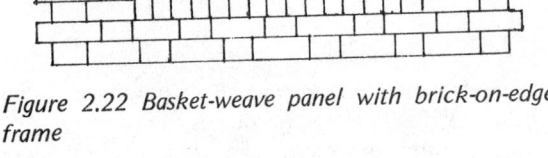
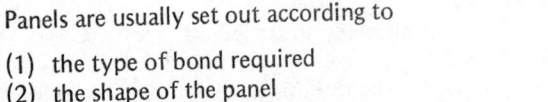
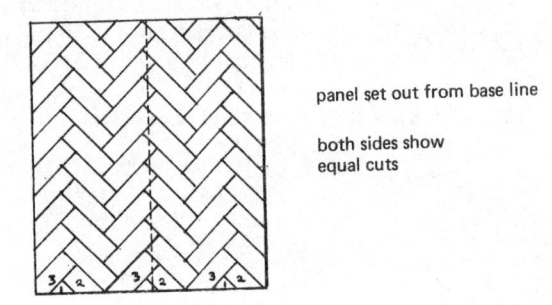

panel set out from base line

both sides show equal cuts

Figure 2.23 Setting out a single herringbone rectangular panel

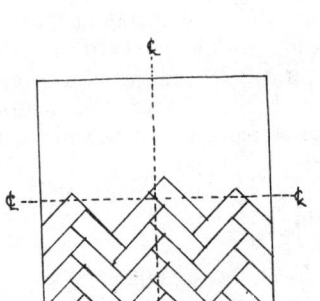

Figure 2.26 Setting out single herringbone in square panel

All herringbone panels, single or double, are set out at 45°. Diagonal lines, vertical and horizontal centre lines are required before setting out can begin (figure 2.24). The same procedure is also used for diagonal basket weave but normal basket weave only requires the vertical and horizontal centre lines.

Marking and Cutting

This has always posed problems for the craftsman, especially when herringbone bonds are used. There are many methods of marking panels for cutting, the simplest being to construct a timber frame to the dimensions of the panel, less the thickness of the mortar joint on all sides (figure 2.27). Nails are positioned in the frame to accommodate the setting-

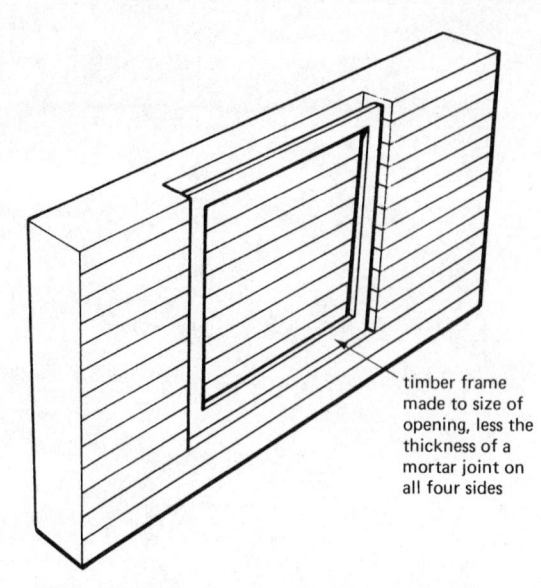

Figure 2.27 Fitting a frame to the panel recess

timber frame
made to size of
opening, less the
thickness of a
mortar joint on
all four sides

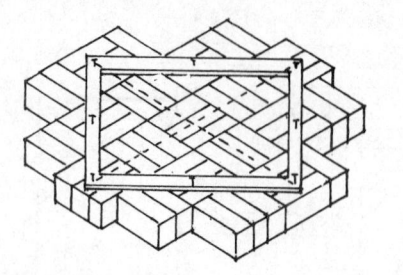

Figure 2.29 Setting out a diagonal basket-weave panel with a templet

out lines, that is, the diagonal, vertical and horizontal centre lines.

The brick panel is then laid out to bond on a flat level surface, with allowances for mortar joints between the bricks. The area should be considerably larger than the dimensions of the panel. The timber frame with the attached lines is then placed on the surface of the bricks, and can be positioned and adjusted to suit the geometrical setting out of the bond (figures 2.28 and 2.29). The perimeter of the panel can then be marked. Whenever necessary the frame can be repositioned and the panel checked for accuracy.

Building-in

The panel can now be built into the recess formed in the wall. Vertical and horizontal working lines are

required: the latter are moved up the face of the walling as the building-in proceeds. Whatever bond is used to form the panel, the work of building-in must start at the base of the panel; with herringbone bonds and also with diagonal basket weave the work starts at the centre of the base line, the first series of bricks being numbered from 1 to 3 or 1 to 4 or 1 to 5 (figures 2.24 and 2.30).

Constant checking is carried out with the use of a small steel setting-out square and all points are positioned by the working lines. Coloured mortar can be used to point the face of the panelling after completion. A slight contrast to the colour of the bricks will obviously enhance the appearance of the finished work.

BONDING PLINTH COURSES

Plinth courses are normally used to reduce the thickness of walls, consequently they are usually seen to occur around the base of walls or possibly at first-floor level. The simplest form of plinth is an offset, where the bricks are set back from the face of the wall (figure 2.31a). With this method no special bricks are

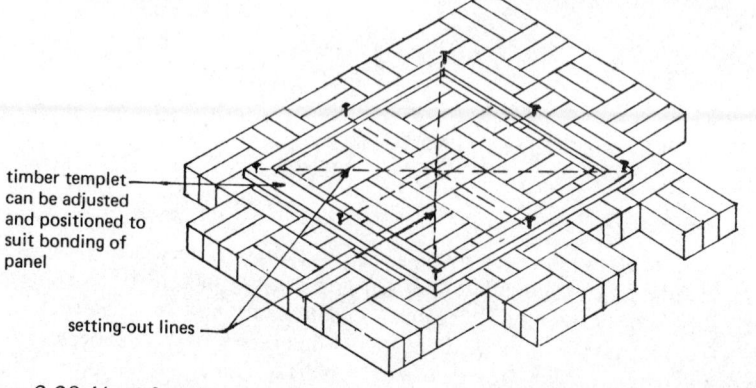

timber templet
can be adjusted
and positioned to
suit bonding of
panel

setting-out lines

Figure 2.28 Use of timber templet for setting out and marking decorative panels

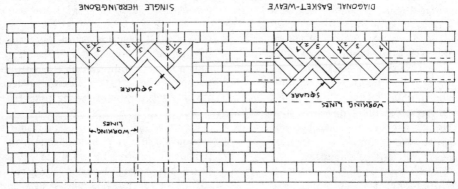

Figure 2.30 Method of building-in panels

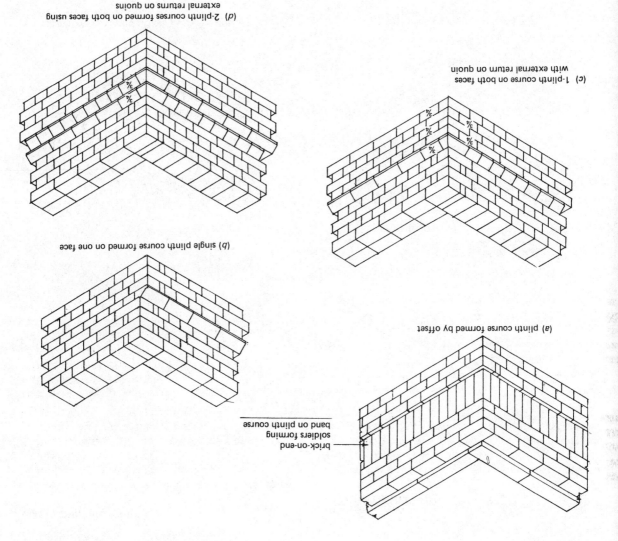

Figure 2.31 Plinth courses formed in English bond

required. If the wall is built in stretcher bond, the width of the cavity is increased below the offset course and the correct width of cavity is formed at the level of the plinth course. For solid walls the collar joint is increased in width to obtain the same effect.

The above method has its disadvantages because the offset formed provides a ledge that is always vulnerable to weather penetration even when a mortar fillet is applied. To provide a better form of weathering, plinth courses are normally formed with purpose-made splay bricks, which can be obtained in headers, stretchers and returns. Bonding these special plinth bricks and plinth courses has always posed considerable problems for the craftsman, but the problems can be greatly reduced by applying the following rules for bonding plinth courses.

Rules for Bonding Plinth Courses

(1) Always consider first, and bond in the three courses of brickwork immediately above the topmost plinth course.
(2) It is permissible to use a queen closer on the face of the wall, at a return quoin, for one course only; this brick is really one of two bevelled closers.
(3) It is acceptable to use header over header or stretcher over stretcher provided that there is a lap of 56 mm (figures 2.32b and 2.33).
(4) If necessary the plinth courses can consist of courses of stretchers, even if this is different from the walling bond.
(5) The courses of brickwork below the bottom plinth course must always be considered last and bonded accordingly. This may involve the use of broken bond, but this must be a secondary consideration, and the inclusion of cut bricks cannot always be avoided (figure 2.31e and 2.32c).

Figure 2.36 illustrates a decorative quoin constructed with plinth courses and inverted splay bricks.

CORBELLING BRICKWORK

It may be necessary during building operations to increase the thickness of walls, or form, or increase the dimensions of attached piers. The operation, called *corbelling*, involves projecting the courses

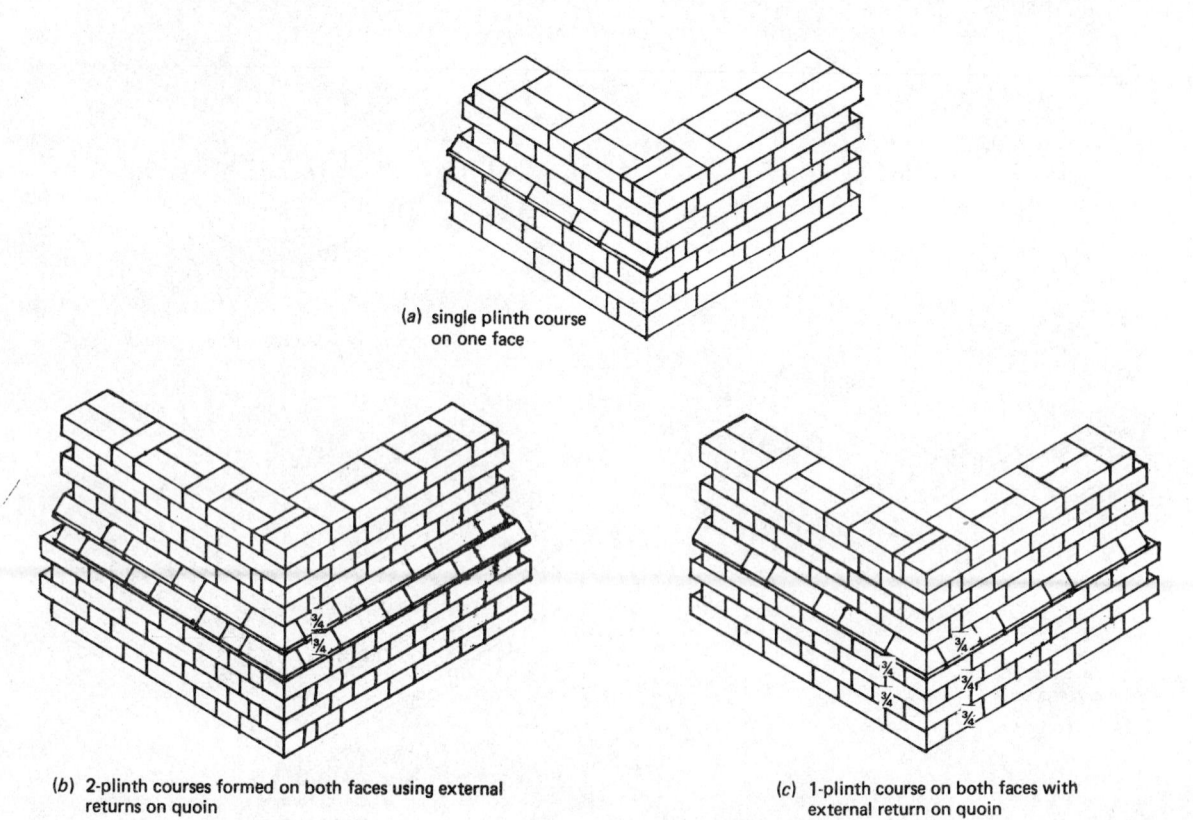

(a) single plinth course
on one face

(b) 2-plinth courses formed on both faces using external
returns on quoin

(c) 1-plinth course on both faces with
external return on quoin

Figure 2.32 Plinth courses formed in Flemish bond

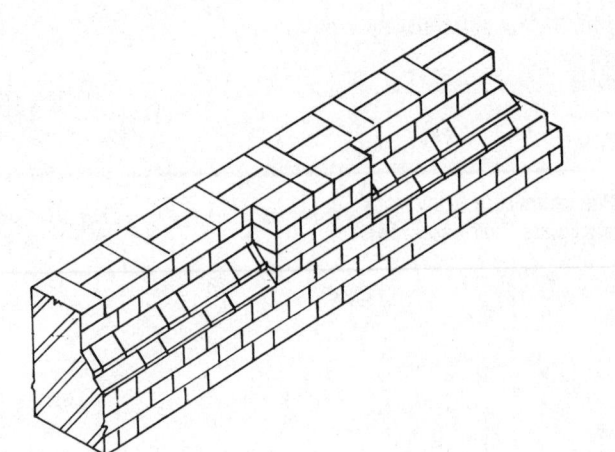

Figure 2.33 Plinth courses used to form an attached pier in English bond

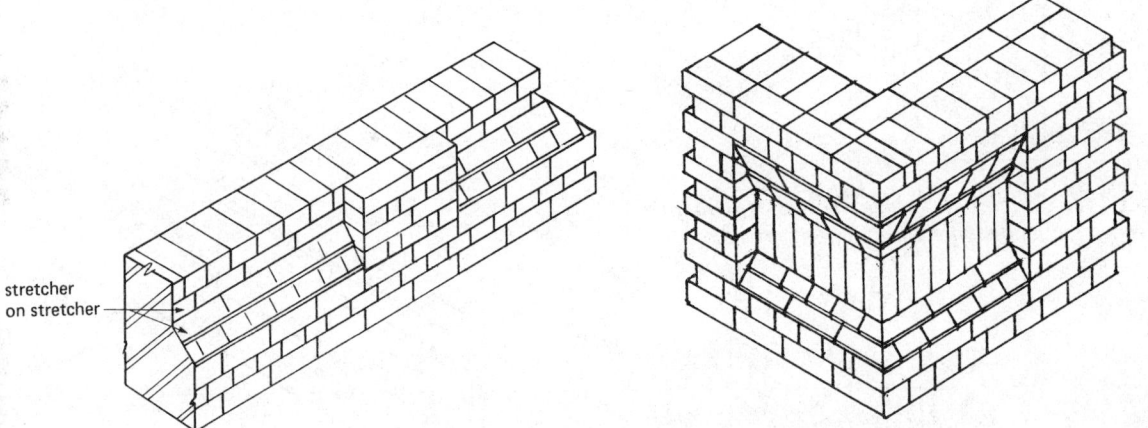

stretcher
on stretcher

Figure 2.34 Plinth courses used to form an attached
pier in Flemish bond

Figure 2.36 Decorative quoin formed with plinth
courses and inverted splay bricks

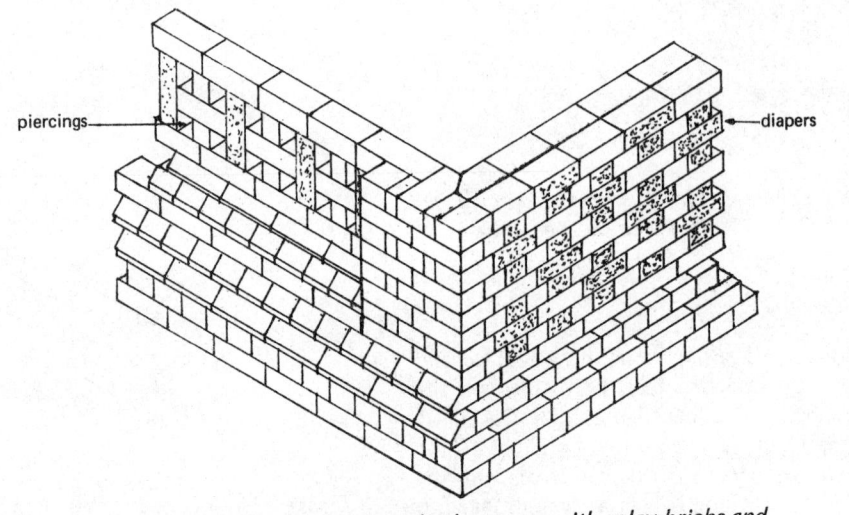

piercings

diapers

Figure 2.35 Method of forming plinth courses with splay bricks and
offsets, incorporating pierced walling and diapers

of brickwork beyond the wall face, and must be carried out with great care.

There are two types of corbel

(1) supported corbels
(2) unsupported corbels.

Supported corbels are projecting courses of brickwork that begin from attached piers (figures 2.37 and 2.38).

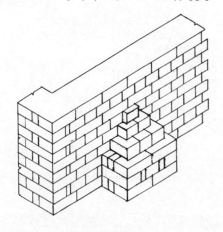

Figure 2.39 Unsupported corbels: forming an attached pier with 56 mm corbels

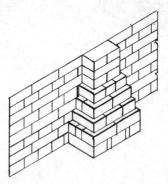

Figure 2.37 Supported corbels: increasing the dimensions of an attached pier with 28 mm corbels

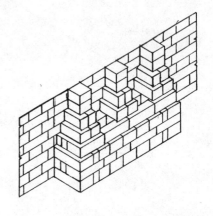

Figure 2.38 Supported corbels: forming an attached pier from half-brick nibs with corbel courses of 28 mm

Unsupported corbels spring from the wall face and are used to form attached piers or to increase the thickness or length of walling (figures 2.39–2.41).

Before corbelling is begun the bonding of the walling above is always considered first and the bonding for the corbelling course is then arranged to coincide with the work above.

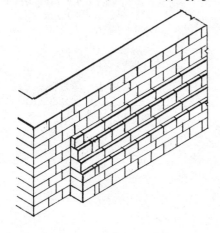

Figure 2.40 Unsupported corbels: increasing the thickness of walls with corbel courses of 28 mm.

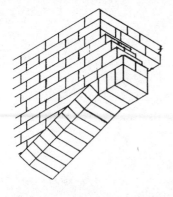

Figure 2.41 Unsupported corbels: a brick knee formed with 38 mm corbels

Rules for Corbelling

(1) All corbel courses should be arranged to project either 28, 38 or 56 mm. The maximum projection is 56 mm.
(2) Corbel courses should be formed in headers whenever possible.
(3) The amount of tie should be 168 mm whenever possible.

Rigid observance of the above rules is obviously not always possible, but tying into the existing wall *is* necessary to obtain stability. Therefore, backweight should always be provided on the tie courses before the next corbel course is begun.

The following recommendations should be observed for corbelling to be successful.

(1) Always use cement mortar for corbelling.
(2) Bricks used for corbels should be only slightly damp and should always be laid with frogs uppermost.
(3) The eye-line should always be formed along the bottom arris of the course (figure 2.42); individual corbels are fixed with the aid of a corbel templet (figure 2.43), while the corbel profile is used to check the completed projections (figure 2.44).

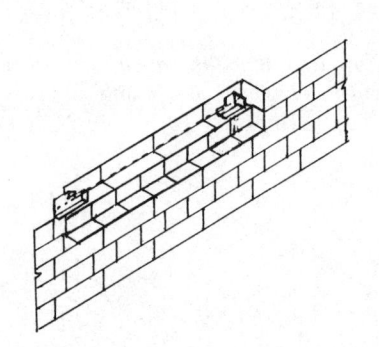

Figure 2.42 Building corbel courses with line and blocks

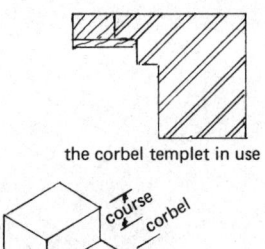

the corbel templet in use

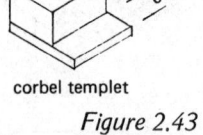

corbel templet

Figure 2.43

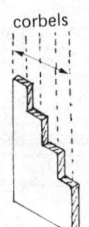

Figure 2.44 Corbel profile board used for checking corbels

TUMBLING-IN

Attached piers and buttresses are often used to provide stability for walls that are subject to lateral pressure. These can be terminated with a simple concrete slab placed on top of the pier, or with courses of splayed bricks (figure 2.45), but where a brickwork finish is required, to provide strength and weathering and to be a decorative feature, the tumbling-in method should be used.

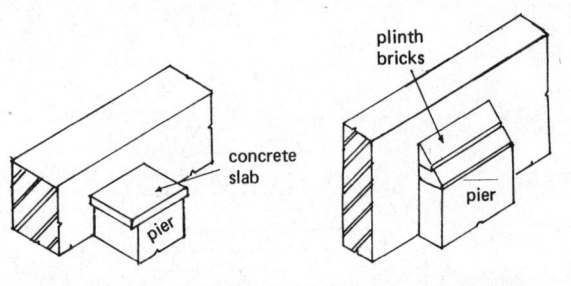

Figure 2.45 Methods of terminating attached piers

To construct tumbling-in brickwork correctly it is necessary to comply with the following rules

(1) where the number of wall courses is odd, the number of tumbled courses should also be odd, but where the number of wall courses is even, the number of tumbled courses should also be even (figures 2.46 and 2.47)
(2) the ratio of tumbled courses to horizontal buttress courses should always be 4:2 or 3:2 (figures 2.48 and 2.50)
(3) the tumbled courses should never enter the wall beyond half the thickness of the wall (figure 2.46)
(4) the bond for the buttress should be continued up the tumbled courses (figures 2.46, 2.49, and 2.50)
(5) a drip should be formed at the start of the tumbled-in work to allow water to fall clear of the buttress face (figure 2.46)

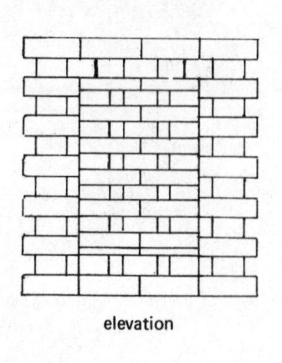

elevation

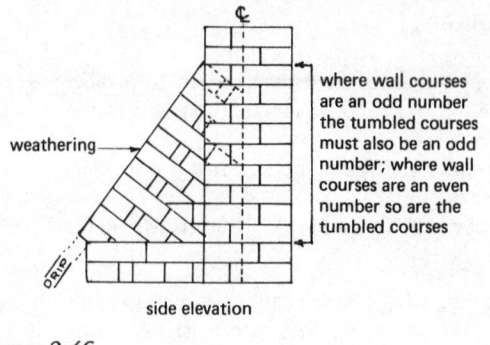

weathering

where wall courses
are an odd number
the tumbled courses
must also be an odd
number; where wall
courses are an even
number so are the
tumbled courses

side elevation

Figure 2.46

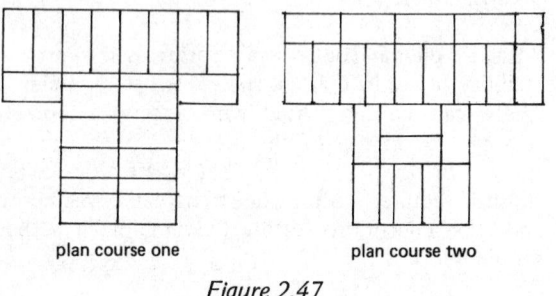

plan course one   plan course two

Figure 2.47

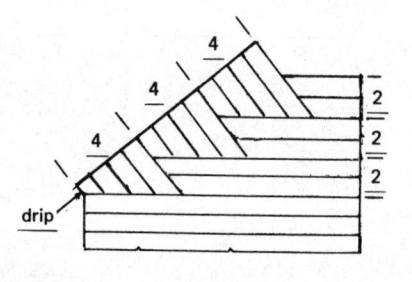

4

4

4

2
2
2

drip

Figure 2.48 Demonstrating the rule that the ratio of
tumbled courses to horizontal courses should always
be 4:2 or 3:2

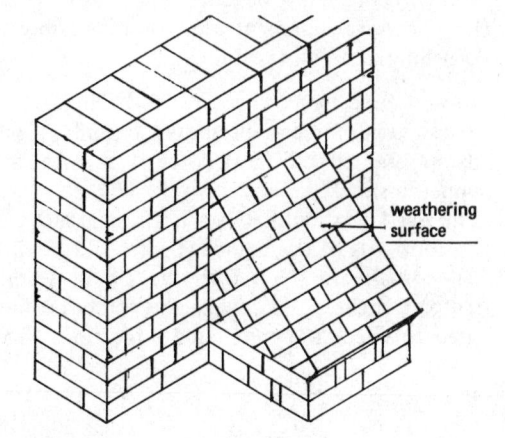

weathering
surface

Figure 2.49 A tumbled-in cap for an attached pier,
formed with an inclined course only

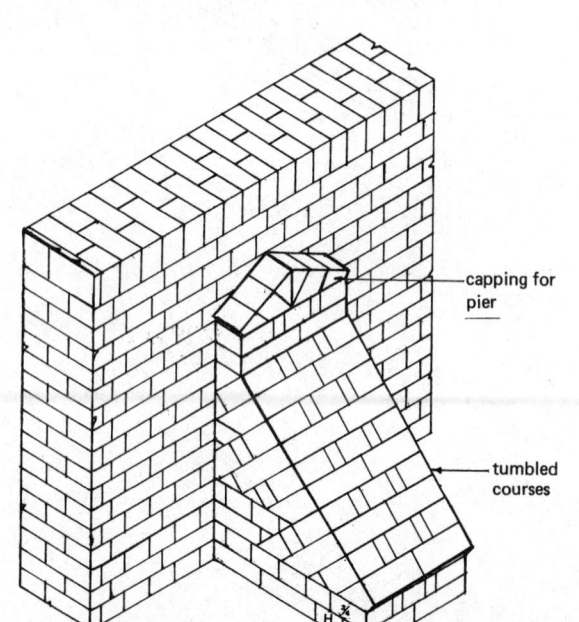

capping for
pier

tumbled
courses

Figure 2.50 Tumbling-in used to reduce the size of
an attached pier

(6) all tumbled courses should be at right-angles to
    the inclination (figures 2.46 and 2.51).
To build tumbled-in brickwork it is necessary to erect
building lines to provide the lines for the inclination;
a gun or stock should also be formed, which can be
used to check the surface of the work and also as a
gauge rod for the tumbled courses (figures 2.51 and
2.52). To comply with rule 1 it may be necessary to
adjust the courses of the tumbled-in work, by either
increasing or reducing the bed joints.
  Checking the tumbled-in courses during erection is
carried out with a steel square to ensure that the
courses are all at right-angles to the inclination (figure
2.51). The angle formed between the tumbled and

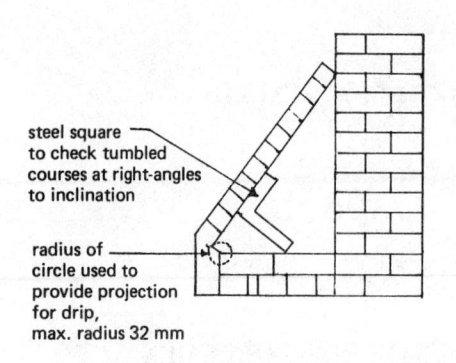

steel square
to check tumbled
courses at right-angles
to inclination

radius of
circle used to
provide projection
for drip,
max. radius 32 mm

Figure 2.51 Building tumbling-in work: method of setting out the tumbling courses

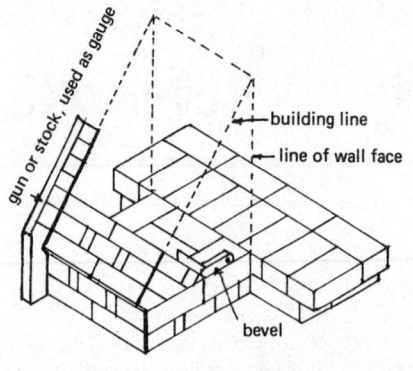

gun or stock, used as gauge

building line

line of wall face

bevel

Figure 2.52 Building tumbled-in work: use of gun or stock to assist in the aligning of tumbled courses, with the bevel providing the angle of cut for the tumbled courses

the horizontal courses is obtained with a bevel, which can then be used to mark the angle of cut (figure 2.52).

To ensure that good effective weathering is provided it is essential that all tumbling-in courses are completely parallel to the angle of inclination, otherwise water may rest on any ledges that are formed. Cement mortar should be used and the bricks should be capable of resisting weather penetration. When tumbling-in work is accurately carried out, the buttress will always be a decorative and functional feature that demonstrates the skills of the bricklayer craftsman.

3
REINFORCED BRICKWORK

The designers of modern buildings are constantly using new techniques, methods and materials to provide structures with increased strength and reductions in loading. Because of this trend the use of reinforcement in walling is becoming a common practice. The reinforcement of brick walls allows for a reduction in wall thickness, and, when walls are reinforced above openings, the compressive strength is increased because the brickwork acts as a beam.

Foundation walling is often reinforced horizontally to prevent settlement, and vertically to resist lateral pressure (figure 3.1). Walls of concrete and brickwork in compound form are constantly used to provide strength and for their decorative qualities. This type of work is often carried out with the inclusion of rod and wire reinforcement (figures 3.6, 3.7 and 3.9).

TYPES OF REINFORCEMENT

Exmet Expanded Metal Mesh

This is a diamond-shaped mesh, obtainable in rolls of 18 m and in widths of 56, 175 and 300 mm. The gauge is 20, 22 or 24. To be effective, the mesh should be completely enveloped within the mortar bed. This type of reinforcement is used in walls and partitions to resist both horizontal and vertical pressure (figure 3.2).

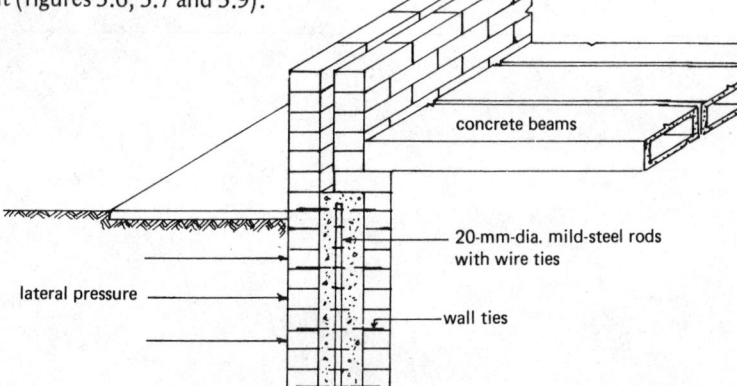

concrete beams

20-mm-dia. mild-steel rods with wire ties

lateral pressure

wall ties

Figure 3.1 Vertical reinforcement in sub-structure walling

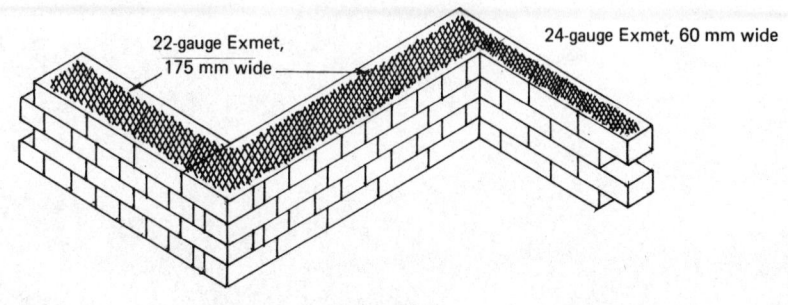

22-gauge Exmet, 175 mm wide

24-gauge Exmet, 60 mm wide

Figure 3.2

B.R.C. Brickforce

This consists of two longitudinal wires which are joined together by spot-welded cross wires that occur every 300 mm. The longitudinal wires are spaced according to the thickness of the walling and rolls are obtainable with these wires at 50, 75 and 150-mm centres. Brickforce is supplied in rolls of 25 m and is used every three to five courses, according to the designer's requirements (see figure 3.3). As with other types of horizontal reinforcement, it should be sealed within the mortar bed and when joining up is required the amount of lap should be at least 225 mm.

Hoop-iron Reinforcement

This is a traditional type of horizontal reinforcement that is still frequently used. The hoop-iron is 25 mm wide and 2 mm thick. It is possible to obtain it in galvanised form or with a bituminous coating. When used in brickwork it should be fixed within the mortar bed, with one band of hoop-iron for each half-brick thickness of walling (figure 3.4). As with other forms of horizontal reinforcement, the hoop-iron should be set at least 25 mm from the face of the walling. Joints are formed with welts or hooks to ensure continuity. Bricktor horizontal reinforcement is shown in figure 3.5

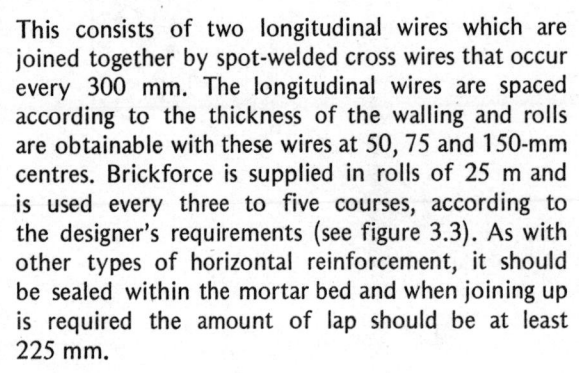

3-mm Brickforce every 3–5 courses

Figure 3.3 Horizontal reinforcement in a prefabricated brick panel

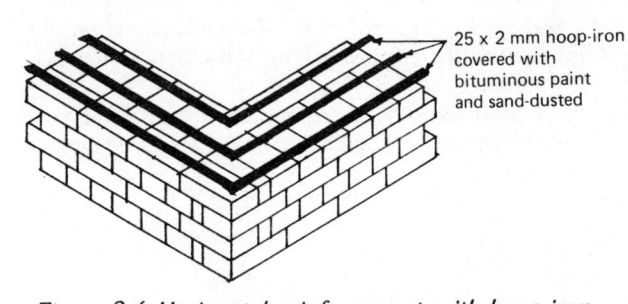

25 x 2 mm hoop-iron covered with bituminous paint and sand-dusted

Figure 3.4 Horizontal reinforcement with hoop-iron

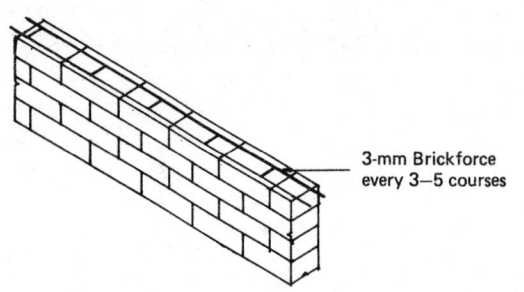

Bricktor using 175 and 65-mm strips

Figure 3.5 Horizontal reinforcement with Bricktor

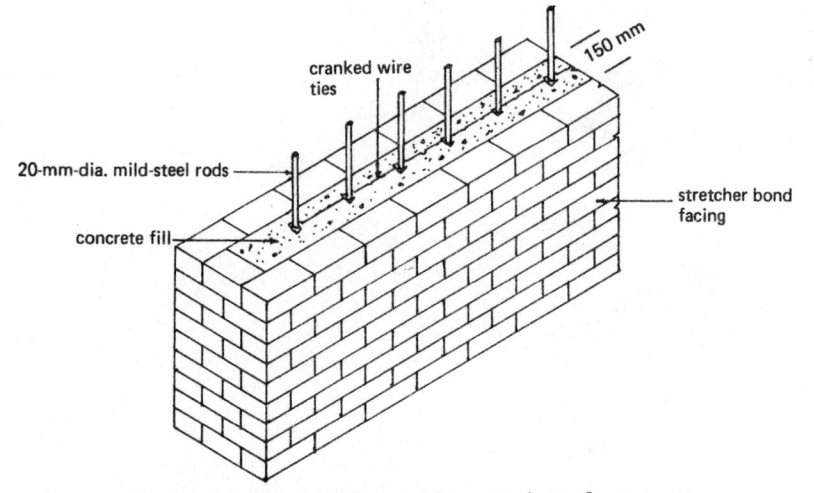

cranked wire ties

150 mm

20-mm-dia. mild-steel rods

concrete fill

stretcher bond facing

Figure 3.6 Thick walling with vertical reinforcement

Rod Reinforcement

This is used vertically to strengthen walls of reduced thickness and to resist lateral stresses. Rods are used with diameters of 12–20 mm, depending on the situation and the requirements of the structural engineer. It is common practice to use this method of reinforcement within concrete pockets formed in the thickness of the wall (figures 3.6 and 3.7) or in concrete walls with brick facings. In these positions wires and stirrups are often used to increase the longitudinal stability of the reinforcement.

In modern construction the use of perforated bricks, with perforations designed to accommodate the vertical rod reinforcement (figure 3.8), provides a wall that is reduced in both thickness and weight but with sufficient strength for the situation in which it is required.

Illustrations of reinforcing with rat-trap bond, wall ties, diagonal bond and brick lintels are given figures 3.9–3.12.

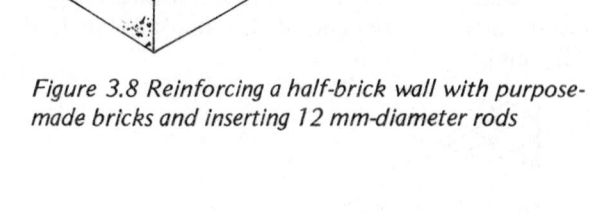

Figure 3.8 Reinforcing a half-brick wall with purpose-made bricks and inserting 12 mm-diameter rods

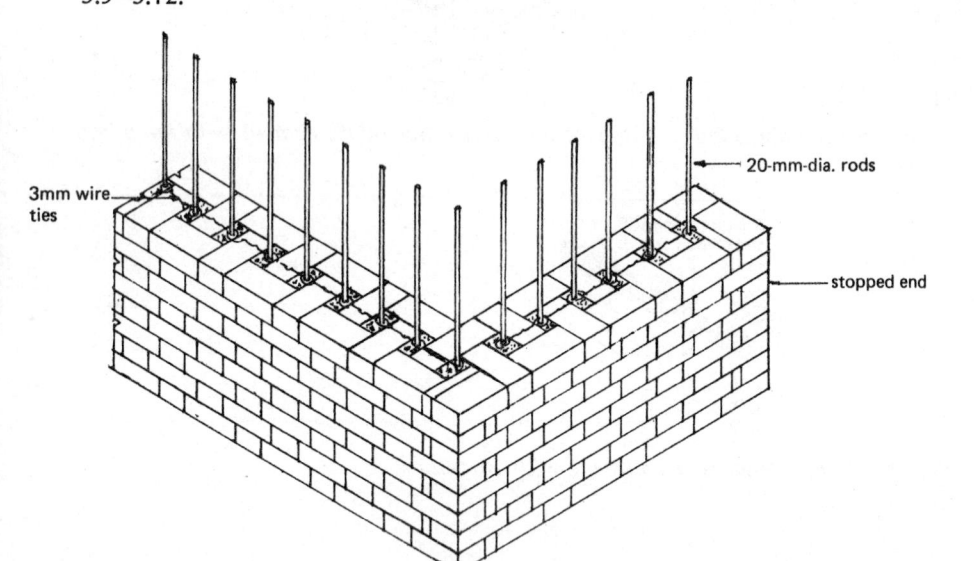

Figure 3.7 Quetta bond with vertical reinforcement

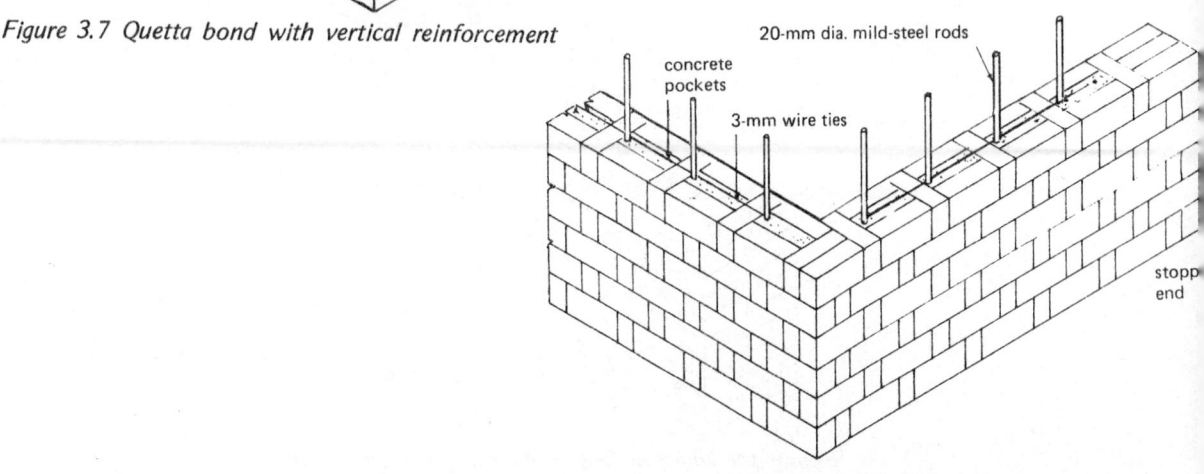

Figure 3.9 Rat-trap bond with vertical reinforcement

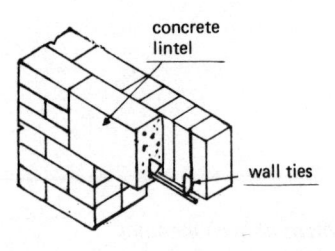

Figure 3.10 *Wall ties fixed on 20 mm rod on every alternate course*

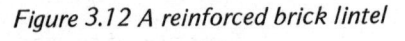

Figure 3.12 *A reinforced brick lintel*

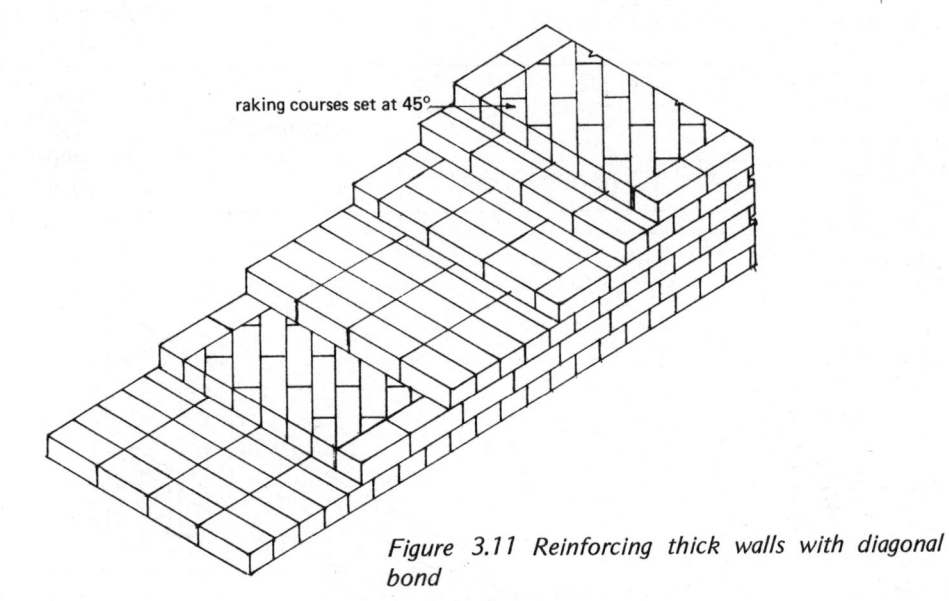

Figure 3.11 *Reinforcing thick walls with diagonal bond*

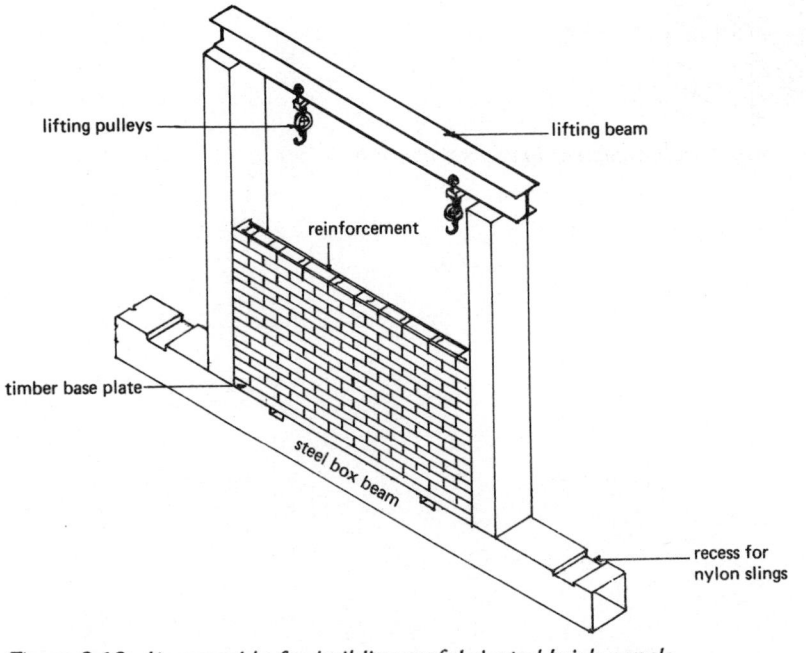

Figure 3.13 *Jig assembly for building prefabricated brick panels*

POSITIONING OF REINFORCEMENT

Recommended positions for the reinforcement, when
walls are to be reinforced above openings, are given in
table 3.1. (See also figure 3.14.)

Table 3.1 Recommended Positions for Reinforcement above Openings

Clear span of opening (m)	Number of courses reinforced	Minimum height of brickwork above opening (mm)
1.2	2	600
1.3	3 (figure 3.14)	750
1.8	3	900
2.0	4	1075

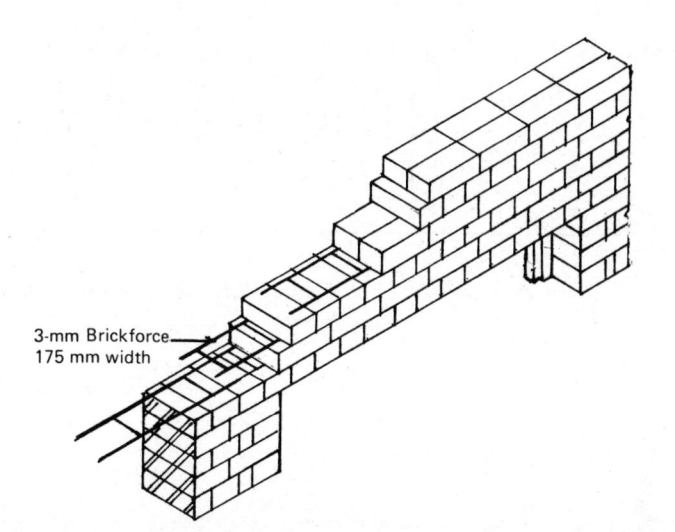

3-mm Brickforce
175 mm width

Figure 3.14 Horizontal reinforcement in courses above openings

4
STEP CONSTRUCTION

Steps are formed to provide access from one level to levels above and should be designed to prevent unnecessary fatigue or physical discomfort. Steps may be formed in brickwork, concrete or a combination of both materials.

To understand the construction of any type of step work it is necessary to have a knowledge of the following definitions

Tread   The upper surface of a step.
Rise   The vertical distance between consecutive steps or between step and landing.
Pitch line   The notional line that connects the nosings of all treads.
Pitch   The angle formed between pitch line and the horizontal.
Nosing   The front edge of a step or tread.
Going   The horizontal distance on the plan between the nosing of a tread and the nosing on the tread above.
Flight   Part of a stairway, or ramp, which may consist of a step or consecutive steps.
Parallel   Steps of uniform width.
Width   The distance between the nosing and the face of the riser (figure 4.1).

REQUIREMENTS FOR BRICK AND CONCRETE STEPS

All steps are now required to be built or formed to comply with the Building Regulations 1976 and the minimum going, pitch, maximum and minimum rise and width are now determined by the type and use of the building.

Steps can be supported by walls or concrete (figures 4.2–4.4). They can be built in during construction or fixed at a later date.

Concrete Steps

When concrete steps are built into the walling, either during construction or later, it is necessary to provide a gauge staff, which should be formed to suit the brickwork courses and the height of the risers (figure 4.2). When the steps are fixed at a later date, the wall should contain recesses, formed in the brickwork or sand courses, to accommodate the concrete steps (figure 4.2).

Concrete steps, when formed as precast units, can have stepped or sloping soffits, the latter having squared seatings at each end to rest on the walling (figures 4.3, 4.5 and 4.6).

Brick Steps

Brick steps should be built of durable bricks, which should be able to resist abrasion and weather and should be set in cement mortar suitable for the type of brick used.

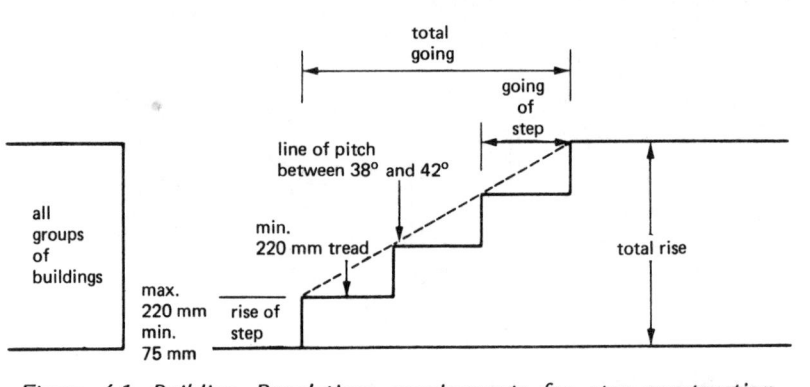

Figure 4.1 Building Regulations requirements for step construction

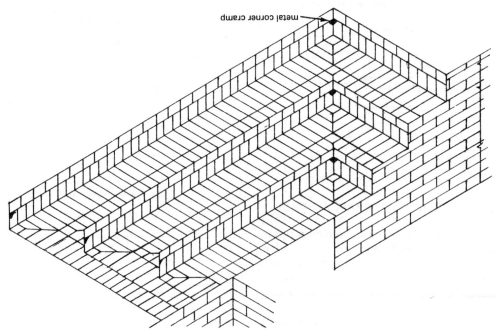

metal corner cramp

Figure 4.8 Pyramidal brick steps

5
DRAINAGE

The object of a drainage system is to convey foul, waste or surface water to the sewer or other place of disposal without danger to health. This means that the pipework must be airtight and watertight in order that both solid matter and liquid matter are removed from a building without foul odours escaping, except where this is part of the design (see page 63).

Definitions of water types are as follows (figure 5.1)

Surface water   The run-off of natural water from the ground surface, including paved areas, roofs and unpaved land.

Ground water   In permeable ground the surface water will percolate downwards towards the water table, being held temporarily in suspension.

Subsoil water   Water occurring naturally below the ground surface, the depth varying with the season.

Waste water   The discharge from lavatory basins, baths, sinks, etc., that is, water not classed as surface water and not contaminated with soil water.

Soil water   The discharge from soil appliances such as water closets, urinals, etc.

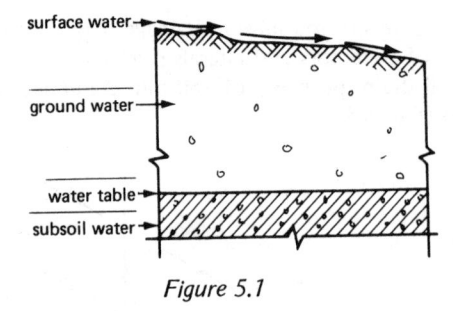

Figure 5.1

SUBSOIL DRAINAGE

Part C2(2) of the Building Regulations states that wherever the dampness or position of the site of a building renders it necessary, the subsoil must be effectively drained as required to protect the building against damage from moisture.

Part C2(3) states that whenever excavation work causes a subsoil drain to be severed, adequate steps must be taken to secure the continued passage of subsoil water through this drain or otherwise to ensure that no subsoil water entering such a drain causes dampness of the site of the building.

Drainage of subsoil water may be necessary for any of the following reasons

(1) to prevent surface flooding and thus improve conditions for building
(2) to lessen the amount of dampness occurring in foundation brickwork
(3) to prevent foundation trenches from becoming waterlogged
(4) to increase the stability of the subsoil and the ground surface
(5) for agricultural purposes
(6) to lessen the humidity that can occur when buildings are erected on damp sites.

Systems

The following are the usual methods of carrying out subsoil drainage, depending on the location of the site and the conditions prevailing.

Natural

Trenches are excavated and pipes laid to follow the natural contours on the site with branch drains discharging into the main drain as necessary (figure 5.2).

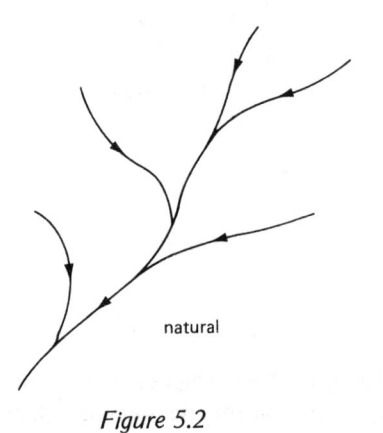

natural

Figure 5.2

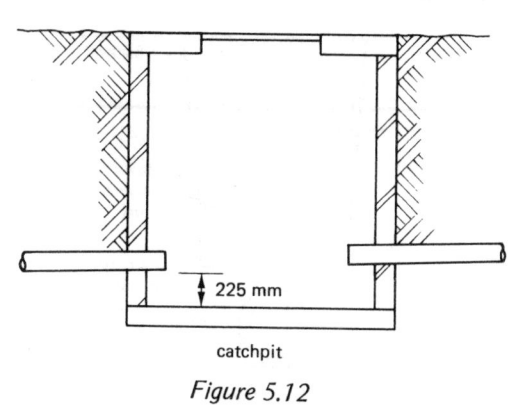

catchpit

Figure 5.12

with large hardcore and covered with a layer of weak concrete. This method is cheap but provides only limited storage space (figure 5.13).

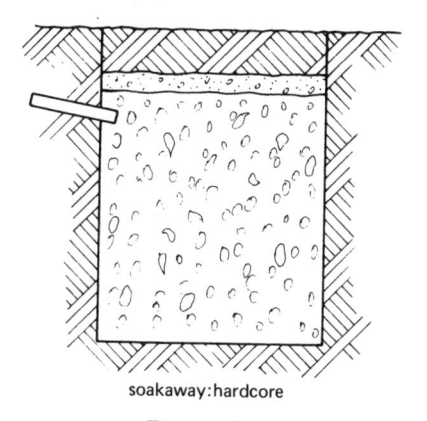

soakaway:hardcore

Figure 5.13

(2)   Precast concrete sections are placed where required and excavation is carried out from the inside, the sections sinking under their own weight. These have a large, easily calculated capacity but costs are increased (figure 5.14).

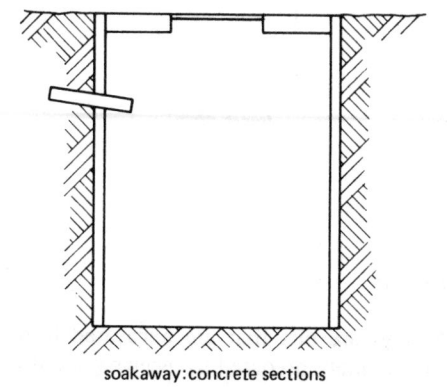

soakaway:concrete sections

Figure 5.14

(3)   An older method is to line the excavated pit with dry brickwork or stone walling, which cuts down on costs, has an easily calculated capacity but is not such a permanent job (figure 5.15).

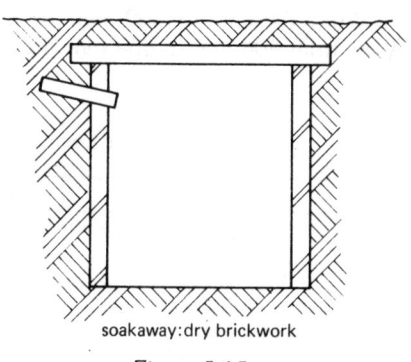

soakaway:dry brickwork

Figure 5.15

Whichever method is decided on it is vital to ascertain that the subsoil is permeable, otherwise the soakaway will become a well. When the nature of the subsoil is not known, trial pits should be excavated on site and the rate of percolation noted. It is often the case that the nature of the subsoil alters with depth and that permeable ground exists below clay, etc. In figure 5.16, for example, it would be virtually useless to dig a soakaway less than 1.5 m deep.

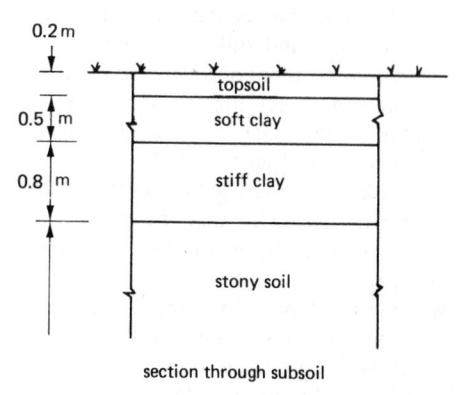

section through subsoil

Figure 5.16

Connection to Waste-water Drains

Any silt being carried along the pipework must not be allowed into a waste-water drain and, therefore, a catchpit is installed before the connection is made. This is a brick chamber similar to an inspection chamber but containing neither channel pipes nor benching (figure 5.12). The outlet is kept about 225 mm above the top of the concrete base and the catchpit must be cleaned out at intervals as required.

Rainwater, Waste and Foul Water

All domestic buildings must be provided with efficient drainage systems in order to dispose of rain, waste and foul water. The system must discharge into a main sewer, a septic tank or a cesspool, depending on the availability of these alternatives. The sewer is laid and maintained by the local authority, usually below the road or footpath, and the house system must be connected to this either at an inspection chamber (see page 64) or between inspection chambers by means of a saddle (see page 56) fitted to the cheek of the sewer.

Septic tanks and cesspools are outside the scope of the craft syllabus but will be covered in an advanced volume.

GENERAL PRINCIPLES OF DRAINAGE

(1) The drainage layout should be as simple and direct as possible.

(2) Materials used should be hard, smooth, non-corrosive and true in shape.

(3) Pipes should be laid to falls to give a self-cleansing velocity. This is generally considered to be a flow of between 0.75 and 3 m/s. While Maguire's rule is to some extent outdated it gives a flow, depending on the type and condition of the pipe, of 1.375 m/s, that is, where a 100 mm pipe is laid at a fall of 1 in 40, a 150 mm pipe is laid at a fall of 1 in 60, and a 225 mm pipe is laid at a fall of 1 in 90.

(4) All joints must be airtight, watertight and free from internal obstruction.

(5) Lines of pipes between inspection chambers are to be as straight as possible, both horizontally and longitudinally.

(6) All inlets to drains must have a water seal of at least 50 mm, except soil and ventilation pipes (see page 63).

(7) Branches should be kept as short as possible.

(8) The greatest volume in flush should be at or near the topmost point of the drain where possible.

(9) All junctions should be made with the flow.

(10) Adequate means of access and inspection must be provided. The Building Regulations 1976 state that inspection chambers shall be placed
(a) at each point where there is such a change of direction or gradient as would prevent any part of the drain being readily cleansed without such a chamber
(b) on a drain or private sewer within 12.5 m from a junction between that drain or private sewer and another, a private or public sewer, unless there is an inspection chamber situated at that junction
(c) at the highest point of a private sewer unless there is a rodding eye at that point
Note The maximum distance between inspection chambers on a drain or private sewer is 90 m.

(11) Pipes must be laid at depths to prevent accidental disturbance or be adequately protected by haunching or be surrounded in concrete.

(12) Waste pipes to ground floors must discharge below grating level but above the water seal.

(13) Vent pipes are to be of sufficient height and to be fitted at the top with a durable cage to prevent ingress of birds, leaves, etc.

(14) Only one connection per dwelling is made to a main sewer (figures 5.28—5.33).

(15) Drains should not pass under buildings unless this is unavoidable, for example, where the length of a drain would be substantially increased or where sufficient fall cannot otherwise be obtained (figure 5.17).

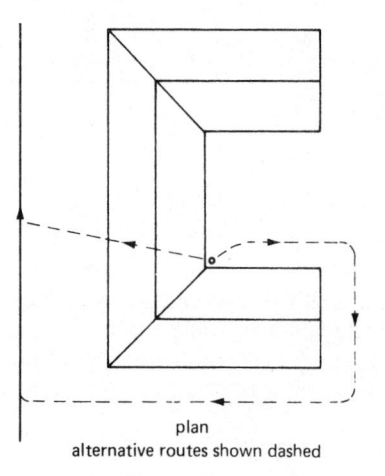

plan
alternative routes shown dashed

Figure 5.17

Where drains do pass under buildings the Building Regulations state that such precautions shall be taken as are necessary to prevent damage to, or loss of water tightness in, the drain or private sewer by differential movement. The principles to be followed here are

(1) Use cast iron pipes, clay pipes surrounded in 150 mm of concrete or flexible joints and/or pipes laid in a granular bed, depending on local authority requirements.

(2) An inspection chamber should be placed outside at least one end of the run under the building.

(3) Where the pipe passes through a wall, either
(a) provide space above the pipe to allow for settlement of the wall
(b) brick round solidly and include at least one flexible joint either side of the wall

Note   Where, for example, an extension to a dwelling is built over an existing drain the local authority usually insists on the drain being exposed and surrounded in concrete that is continued up to the level of the natural foundation. This may not be necessary in the case of a deep drain which is usually left undisturbed. For drains that are laid close to foundations the regulations are shown in figure 5.18.

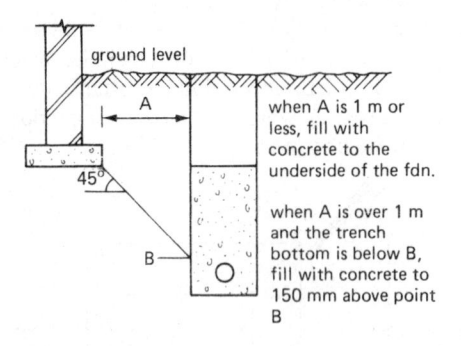

ground level

A

when A is 1 m or less, fill with concrete to the underside of the fdn.

45°

B

when A is over 1 m and the trench bottom is below B, fill with concrete to 150 mm above point B

Figure 5.18

Drainpipes

Pipes for drainage systems are manufactured from many materials and, while the use of cast iron has been spoken of, the craft certificate syllabus mentions only those of vitrified clay and pitch fibre.

Vitrified Clay Pipes

British Standards 65 and 540 Part 1 specify the requirements for clay pipes and fittings, with or without sockets, that are suitable for drains and sewers under two descriptions

(1) British Standard, for foul sewage and/or surface water
(2) British Standard Surface Water, for surface water only

and pipes must be clearly marked as such. Pipes and fittings of either description may be glazed or unglazed externally, internally or both.

Part 2 of the Standard specifies the requirements for flexible joints for use with pipes complying with Part 1.

Pipes with Sockets   These have rigid or flexible joints and, while the latter cost more to purchase, they have many advantages over the more traditional rigid-jointed pipes

(1) Pipelaying is much quicker with the simple push-fit joints; thus labour costs are reduced.
(2) The joints, once made, are immediately watertight; thus testing is not held up.
(3) Joints can be made in waterlogged trenches or freezing conditions.
(4) The pipes are self-centring; mis-alignment at the joints cannot occur.
(5) The fact that the joints are flexible allows for slight distortions of the pipeline due to ground movement without loss of watertightness.

Two types of flexible joint, the O-ring joint and the polyurethane joint, suitable for socketed pipes, are shown in figure 5.19*c* and *d*, the sleeve joint (*e*) being for butt-ended pipes only.

Two rigid joints are shown in figure 5.19*a* and *b*, the gaskin/cement joint being the most common, and this is described later. A complete range of fittings is available for both rigid and flexible-jointed pipes, some of which are shown in figures 5.20 and 5.21.

When ordering channel pipes it is necessary to stipulate whether a left or right-hand fitting is required. For example, a left-hand fitting, when viewed against the direction of flow, branches or bends to the left; and where a double spaced junction is required, the first branch from the spigot is stated first, that is, left/right; right/left (figure 5.21).

Pipes without Sockets   In the Hep-sleve system butt-ended pipes are jointed with a polypropylene coupling and sealing ring (sleeve joint, figure 5.19*e*).

After a check has been made that the pipe end and the coupling are clean, the pipe is placed vertically on a clean, hard surface and the top end is lubricated. A little downward pressure on the coupling easily forces it into place against the central stop. The first pipe in the trench should be butted up to a stopboard and the lubricated end of the next pipe is forced into the collar of the first. Little waste occurs with this system since cut pipes can be utilised after trimming the cut end as necessary. As with collared pipes, a wide range of fittings is available.

Pitch-fibre Pipes

This type of pipe is flexible, light in weight and generally available in lengths between 2400 and 3040 mm (figure 5.22). While pitch-fibre pipes are considered unsuitable for continuously running hot water

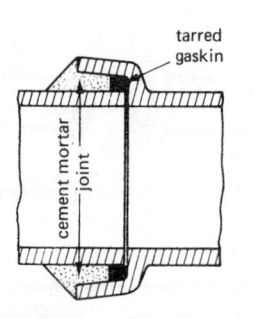

(a) tarred gaskin and cement mortar joint

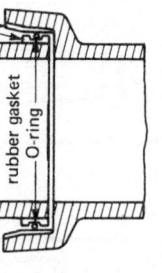

(c) O-ring joint

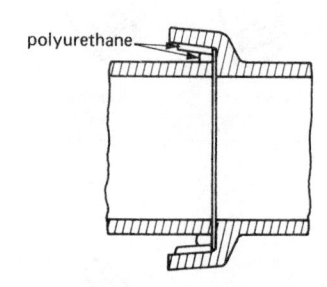

(e) sleeve joint

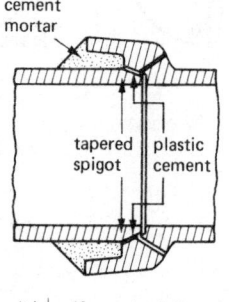

(b) self-centring joint

(d) polyurethane joint

Figure 5.19 Rigid and flexible joints

such as the discharge from laundries, or for wastes containing pitch solvents (petrol, oil and fat), they have the following advantages over rigid-jointed clay pipes

(1) their flexibility makes them suitable for ground liable to differential settlement
(2) concrete bedding, haunching and/or surround is very rarely required
(3) there are no waste pieces, and cut lengths can be re-used
(4) pipelaying can continue in freezing weather and waterlogged trenches
(5) immediate testing on completion is possible
(6) the pipes can be laid at lower gradients than pipes of certain other materials, for example, 100 mm pipe is laid at a gradient of 1 in 85.

Trench Excavation

It is important not to open trenches too far in advance of pipelaying to reduce the possibility of side collapse in loose ground. Where trenches are wide, usually because of considerable depth, the normal practice is to excavate to the position of the crown of the pipe and dig a sub-trench to take the pipe (figure 5.23). This process cuts down the pressure of the backfill on the pipeline.

The sub-trench should be dug at least 100 mm below the proposed invert level for the bedding material, and the trench bottom must be of even bearing. This necessitates removing any large stones and filling any holes so formed, including any existing soft spots, with extra bedding material.

Bedding

Where the subsoil is suitable, for example, where it is free draining coarse sand or loam, the practice is to replace the soil evenly, consolidate it, and bed the pipes down on this, but granular materials are mostly used for this purpose, for example, pea gravel or broken stone of maximum size 19 mm. Clay and chalk, which are affected by percolating water, should not be used for either bedding or sidefilling, and bricks, etc. must never be used to pack pipes into line.

Laying the Pipes

The pipes are laid directly on the bedding material and the joints are made with either a straight (sleeve) coupling or a snap-ring joint (figures 5.24 and 5.27).

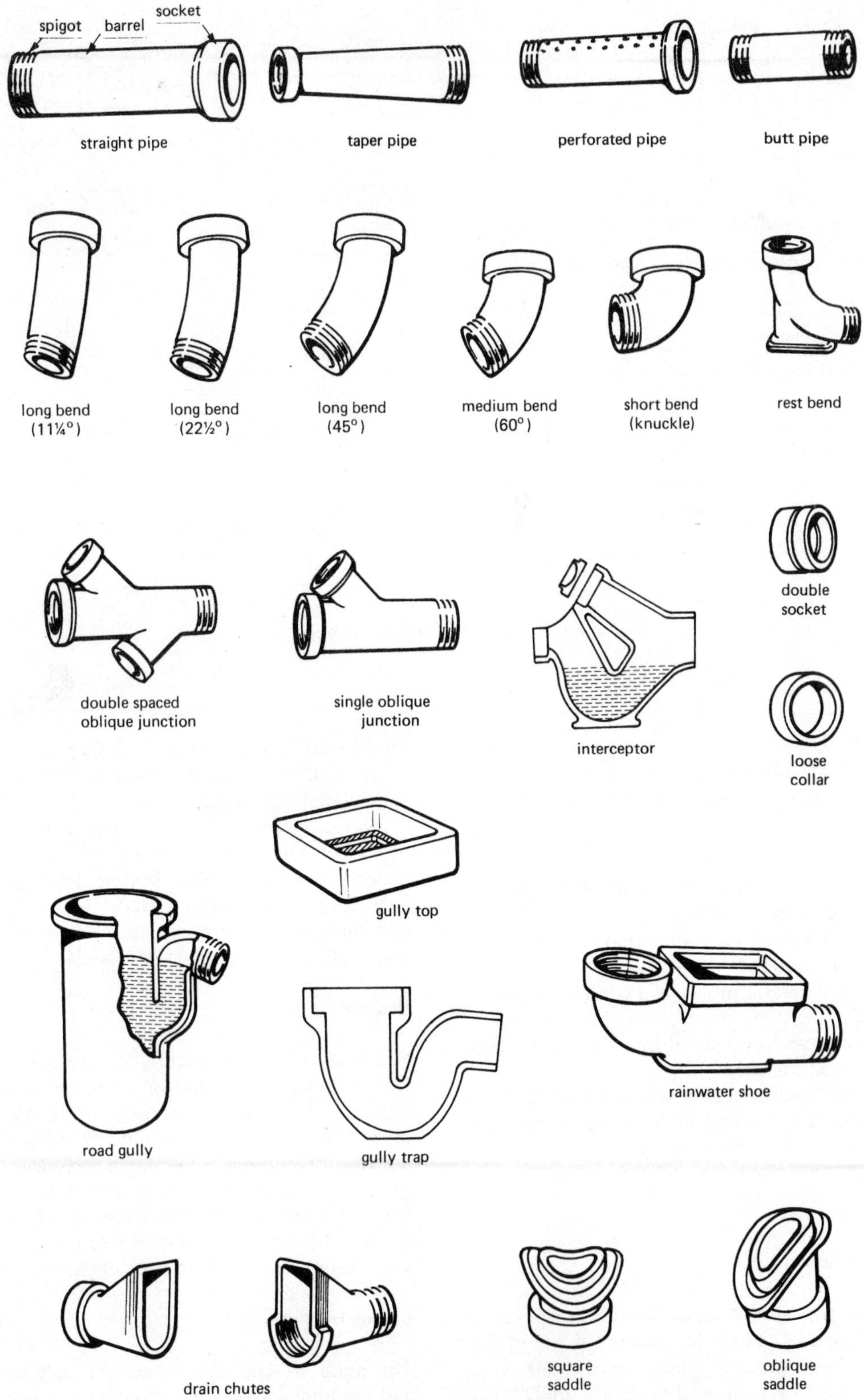

spigot barrel socket

straight pipe   taper pipe   perforated pipe   butt pipe

long bend   long bend   long bend   medium bend   short bend   rest bend
(11¼°)   (22½°)   (45°)   (60°)   (knuckle)

double spaced   single oblique   interceptor   double
oblique junction   junction   socket

loose
collar

gully top

road gully   gully trap   rainwater shoe

drain chutes   square   oblique
                      saddle   saddle

Figure 5.20 Drainpipes and fittings

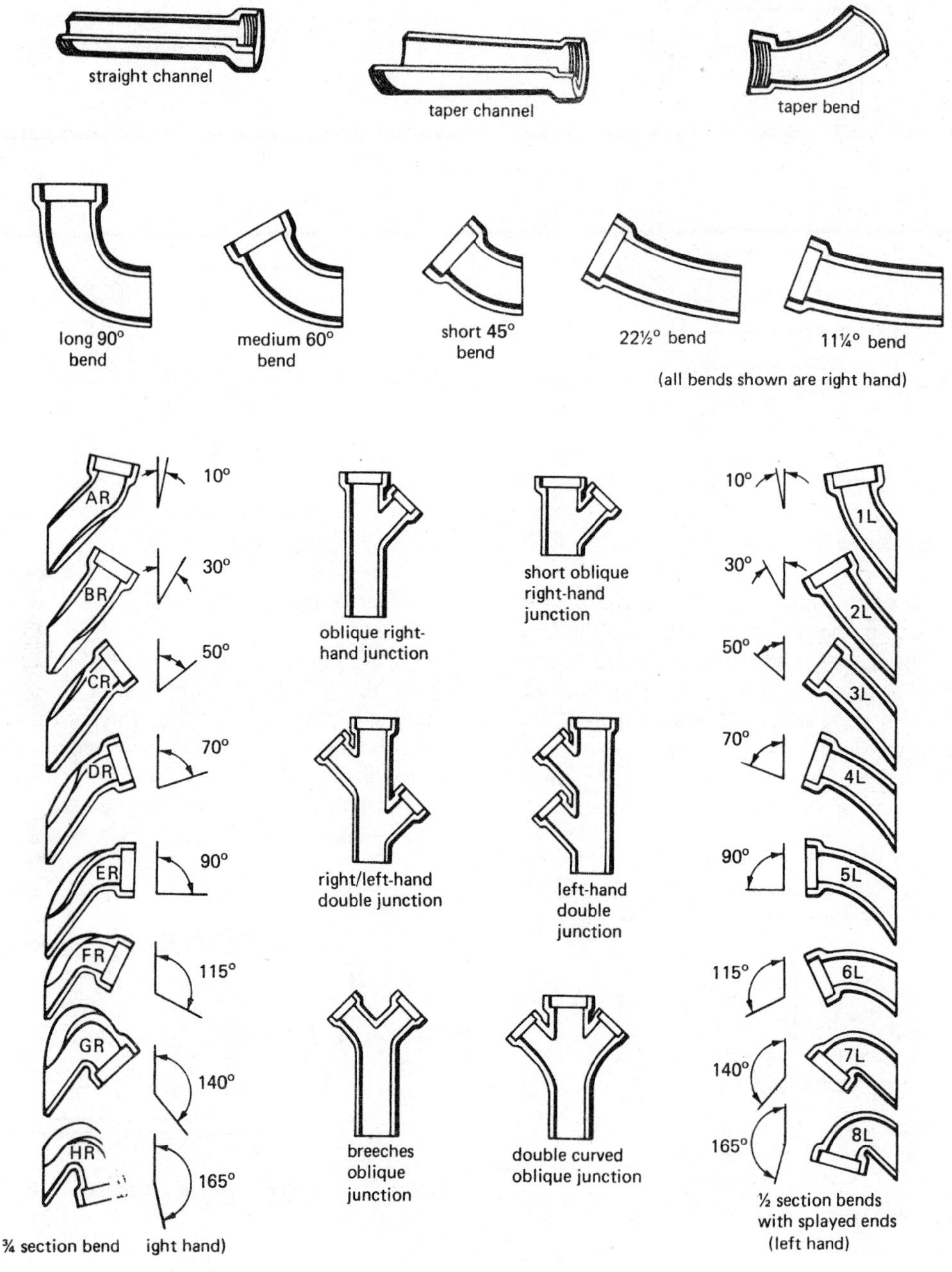

straight channel

taper channel

taper bend

long 90°
bend

medium 60°
bend

short 45°
bend

22½° bend

11¼° bend

(all bends shown are right hand)

AR 10°

BR 30°

CR 50°

DR 70°

ER 90°

FR 115°

GR 140°

HR 165°

¾ section bend (right hand)

oblique right-
hand junction

short oblique
right-hand
junction

right/left-hand
double junction

left-hand
double
junction

breeches
oblique
junction

double curved
oblique junction

10° 1L

30° 2L

50° 3L

70° 4L

90° 5L

115° 6L

140° 7L

165° 8L

½ section bends
with splayed ends
(left hand)

Figure 5.21 Channel pipes

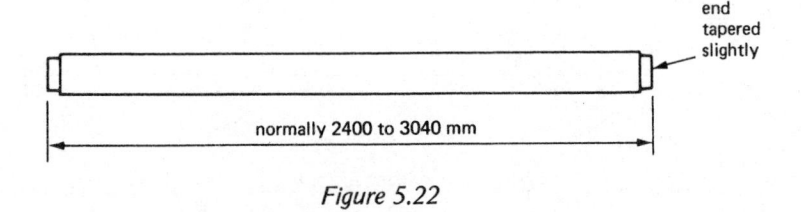

end
tapered
slightly

normally 2400 to 3040 mm

Figure 5.22

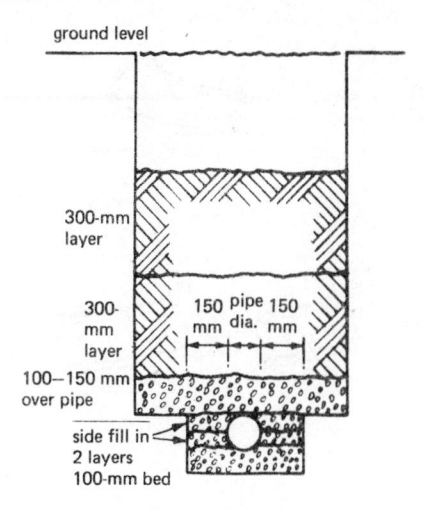

Figure 5.23

Straight Couplings (figure 5.24)

The pipe spigots are wiped clean and a coupling is fitted hand-tight to one spigot of each. Place the first pipe in position against a suitable stopboard (figure 5.25) and fit the spigot of the second pipe into the coupling of the first. A softwood dolly (figure 5.25) is placed against the coupling of the second pipe, which is then driven home with a 1—2 kg hammer. The amount of drive should be 6—7 mm and it is

important not to overdrive. It is possible to complete a considerable length of pipe at ground level if required and to lower it into the trench afterwards.

Note   When using these couplings it is necessary to support the back of a bend or a branch junction while driving (figure 5.26).

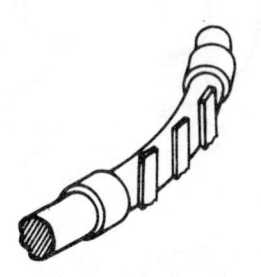

Figure 5.26 Pipes supported at bends during driving

Snap-ring Couplings (figure 5.27)

The snap-ring is placed over the end of a plain-ended pipe with the flat of the ring against the pipe, and the coupling is pushed on to this, forcing the ring to roll along the pipe. The ring is compressed and jumps into the required position. Pipes are jointed in a trench by pushing a pipe with a coupling fitted into the coupling on a previously laid pipe using a spade as a lever.

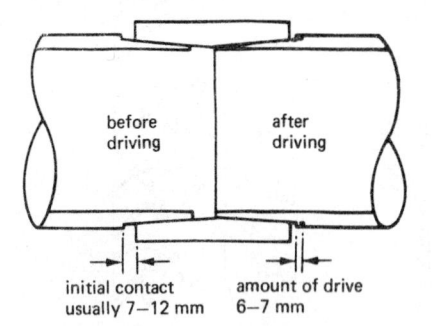

Figure 5.24 Straight coupling for tapered pipes

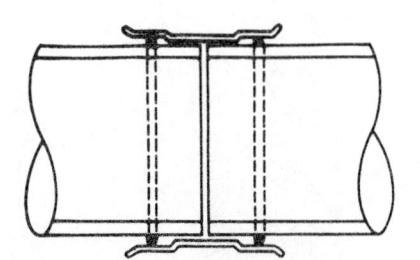

Figure 5.27 Polypropylene snap-ring joint for plain-ended pipes

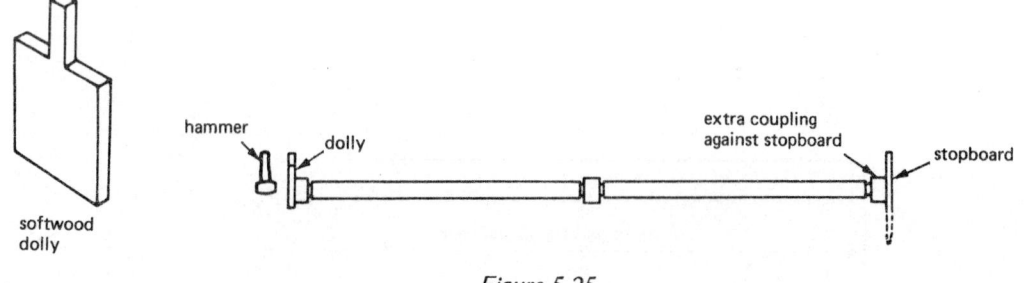

Figure 5.25

Side and Backfilling (figure 5.23)

Further bedding material should be carefully compacted around the pipes in 75 mm layers to the crown
of the pipe, after which inspection and testing should
be carried out. A further 100–150 mm of bedding is
placed on top of the pipe, then backfilling, with
excavated material, should be completed in 300 mm
layers. No mechanical rammers must be used until
backfilling is at least 300 mm above the crown.

Where pipes are laid close to ground level, narrow
slabs of precast concrete should be laid above the
pipes on a cushion of bedding material.

DRAINAGE SYSTEMS

There are three systems of drainage, the combined,
the separate and the part separate system and it is
important to understand at this point that only one
of these systems is used in each town or district.

Combined System (figure 5.28)

In the combined system one large system takes the
discharge from top, waste and foul-water fittings.
This system is found mainly in Scotland and some
coastal areas where sea outfalls are used.

Advantages are

(1) the pipe layout on site is usually simple and
    straightforward
(2) there is no chance of connecting to the wrong
    sewer
(3) there is only one set of pipes and sewer to lay
    and maintain.
    Disadvantages are

(1) the combined sewer and the treatment works
    must be large or they will not be adequate in wet
    weather
(2) storm overflows may be necessary to divert any
    overload to storage tanks, streams, etc., and
    untreated sewage may cause pollution
(3) both foul and surface water need to be treated at
    the sewage works.

Separate System (figure 5.29)

With this system two completely separate sets of
pipes lead to two different sewers, one conveying foul
and waste water from sinks, baths, W.C.s, bidets, etc.,
the other taking rainwater from downspouts, paved
areas, etc.

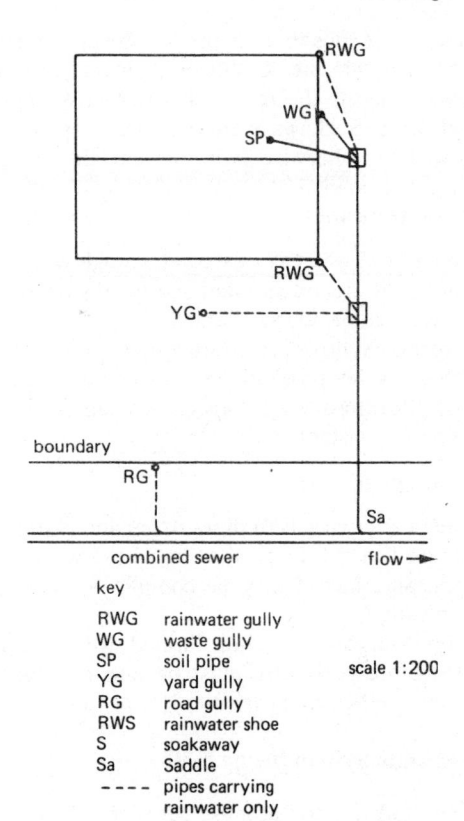

key

RWG	rainwater gully
WG	waste gully
SP	soil pipe
YG	yard gully
RG	road gully
RWS	rainwater shoe
S	soakaway
Sa	Saddle
- - - -	pipes carrying rainwater only

scale 1:200

Figure 5.28 The combined system

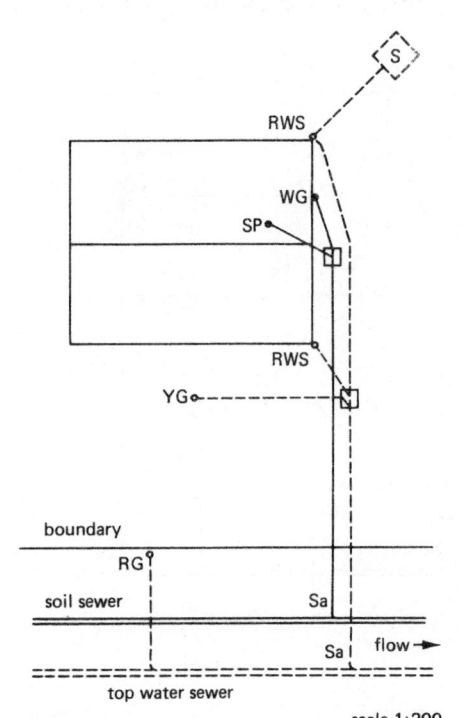

scale 1:200.

*Figure 5.29 The separate system (key as for figure
5.28)*

It may be possible to use soakaways where the subsoil is permeable, as shown in figure 5.29, where the soakaway at the rear is an alternative to a long run of pipes. Soakaways are placed at least 3 m from the dwelling, otherwise drain runs are laid out for the best result possible.

Advantages are

(1) only foul water is treated at sewage works, thus both the treatment plant and the diameter of the sewer can be relatively small
(2) storm overflows are not required
(3) there is no possibility of water pollution from overflowing sewage during bad weather, since the flow is constant.

Disadvantages are

(1) there are two sets of pipes to lay and maintain
(2) there is a risk of connecting to the wrong sewer
(3) the pipe layout may be complicated with pipes crossing
(4) the foul sewer is not flushed with rainwater, therefore great care must be taken to ascertain that a self-cleaning velocity is kept to throughout.

Part-separate System (figure 5.30)

This method is a compromise between the combined system and the separate system. One sewer deals with street gullies and as much roof water as possible, the other takes foul and waste water and a small amount of rainwater, preferably, for example, via a gully at the top of the system, which will flush the drain as it flows during wet weather.

Advantages are

(1) the layout is usually easier and cheaper than for the separate system
(2) foul drains are flushed in rainy periods.

Disadvantages are

(1) there are two drains and two sewers to lay and maintain
(2) there are two connections to be made to the sewers
(3) the layout is costlier and more difficult than for the combined system.

Conclusions are that the combined system, despite its advantages, is considered the worst system, the separate system and the part-separate system being preferred.

Note   The provision and connection of the road gully in all the systems is the responsibility of the local authority.

Measurement of Pipes

The simplest method is to measure the complete length of pipeline and specify all fittings.

Example 5.1. The Combined System (figure 5.28)

Straight Pipes

Top inspection chamber (IC) to sewer	14.6 m
Rainwater gully (RWG) to top IC	4.0
Waste gully (WG) to top IC	1.8
Soil pipe (SP) to top IC	3.2
RWG to lower IC	2.0
Yard gully (YG) to lower IC	6.0
	31.6 m

Fittings

2 rainwater gullies (to take fall pipes)

1 rest bend (foot of soil pipe)

1 waste gully (sink waste)

1 yard gully (surface water from paved areas)

1 saddle (connection to main)

1 slow bend (into top IC from RWG)

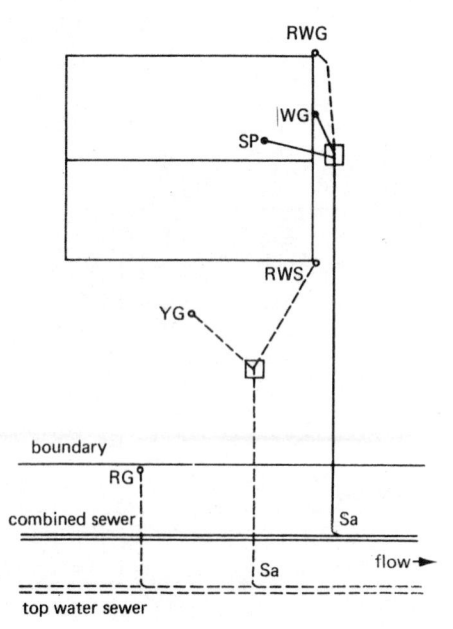

scale 1:200

Figure 5.30 The part-separate system (key as for figure 5.28)

Channel Fittings

Top IC

 1 left-hand double oblique junction

 1 left-hand bend

Lower IC

 1 left-hand double oblique junction

 1 left-hand slow bend

Example 5.2. The Separate System (figure 5.29)
 (assuming soakaway impracticable)

Straight Pipes

Soil length

Top IC to main	14.6 m
WG to IC	2.2
SP to IC	2.6
	19.4 m (British Standard)

Surface water

Top RWS to main	21.0
RWS to IC	2.0
YG to IC	8.0
	31.0 m (British Standard Surface Water)

Alternatively, 50.4 m of British Standard straight pipes could be ordered.

Fittings

 2 rainwater shoes (gullies not necessary)

 1 waste gully

 1 rest bend

 1 yard gully

 2 slow bends (for use in long length)

 2 saddles (two connections)

 1 slow bend (top RWS to IC)

Channel Fittings

Top IC

 1 single oblique left-hand junction

 1 left-hand bend

Lower IC

 1 double oblique left-hand junction

 2 left-hand bends

Example 5.3. The Part-separate System (figure 5.30)

Straight Pipes

Top RWG to main	19.0
WG to top IC	1.2
SP to top IC	2.6
Lower IC to main	9.0
RWS to lower IC	4.4
YG to lower IC	2.8
	39.0 m

Fittings

 1 rainwater gully (connects to soil sewer)

 1 rainwater shoe (connects to surface-water sewer)

 1 rest bend

 1 waste gully

 1 yard gully

 2 saddles

Channel Fittings

Top IC

 1 double oblique left-hand junction

 1 left-hand bend

Lower IC

 1 breeches oblique junction

Note To give a complete specification for channel fittings required for the inspection chambers, they would have to be drawn to a larger scale, as in figure 5.36, for example.

 Figures 5.31–5.33 show one further example of each of the three drainage systems.

Note It may be decided that the lower rainwater gully in the part-separate system (figure 5.33) should discharge into the top-water sewer, in which case the pipes would cross.

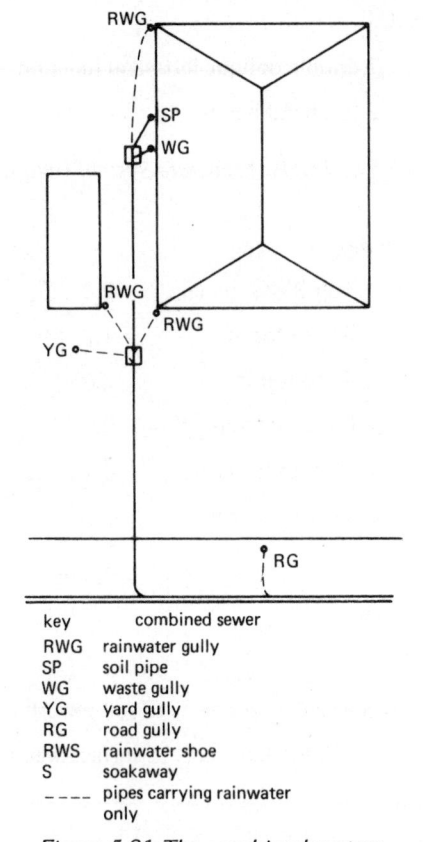

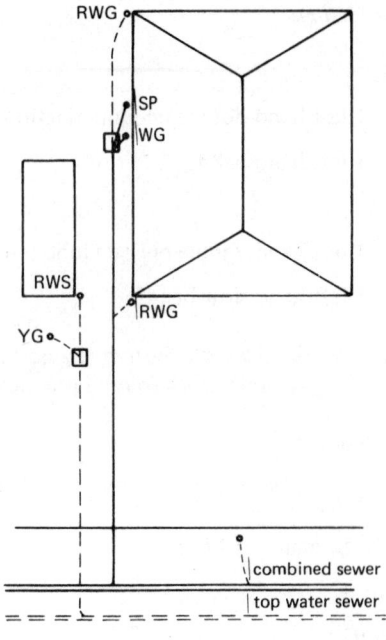

key   combined sewer
RWG   rainwater gully
SP   soil pipe
WG   waste gully
YG   yard gully
RG   road gully
RWS   rainwater shoe
S   soakaway
----   pipes carrying rainwater
          only

Figure 5.31 The combined system

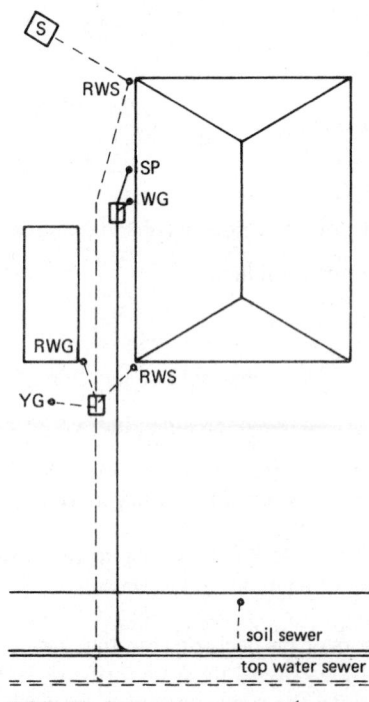

Figure 5.32 The separate system (key as for figure 5.31)

Figure 5.33 The part-separate system (key as for figure 5.31)

VENTILATION OF DRAINS

A free circulation of air must be provided through the pipes forming a domestic drainage system. This is accomplished in one of two ways, depending on the requirements of the local authority.

Ventilating without an Interceptor Trap (figure 5.34)

This is generally considered to be the best method, every drain ventilating the main sewer and thus preventing the build-up of sewer gases.

Ventilating with an Interceptor Trap (figure 5.35)

With this method, sewer gases are prevented from entering a private sewer by means of the water seal in the interceptor trap. Through ventilation of the domestic system is achieved by the provision of a fresh-air inlet at the intercepting chamber, which is installed just inside the site boundary. Fresh air is drawn into the fresh air inlet via a one-way flap, passes through the system, and is released from the top of the vent pipe, which should be provided at or near the top of the drainage system.

Disadvantages of this system include
(1) the installation of an intercepting chamber increases the cost of the system

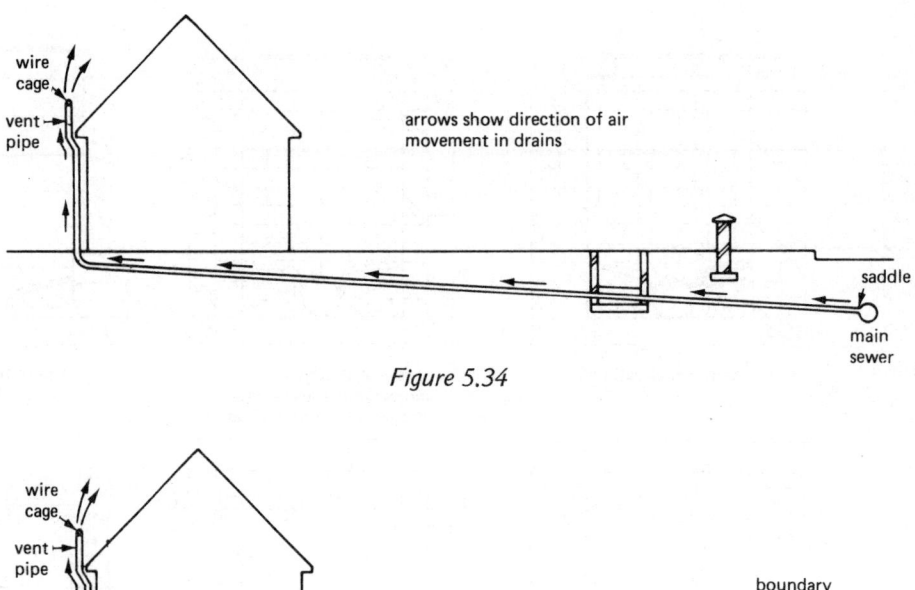

Figure 5.34

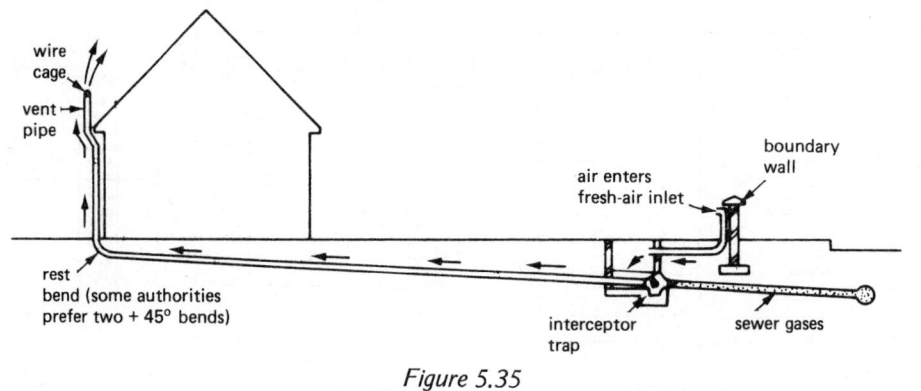

Figure 5.35

(2) the trap itself is liable to become blocked and may require regular cleaning
(3) if the fresh-air inlet suffers damage or becomes faulty, ventilation ceases to take place
(4) some other means of preventing the build-up of gases within the sewer must be provided.

INSPECTION CHAMBERS

The purpose of an inspection chamber is to provide access for inspection and cleansing. Inspection chambers are constructed from the following materials

(1) class B engineering bricks
(2) precast concrete sections surrounded in concrete 100–150 mm thick
(3) *in-situ* concrete
(4) for surface-water drains, good quality bricks, rendered externally where deemed necessary
(5) glass-reinforced plastic.

The following notes are relevant to the construction of inspection chambers.

(1) The concrete base must be at least 100 mm thick, the thickness increasing with depth.

(2) The base can be of the same length and breadth as the overall plan area of the chamber, that is, no spread is required.
(3) The internal size varies with the depth and the number of branch drains entering: 600 x 450 mm is the minimum, increasing to 1350 x 1130 mm for an extra deep manhole.
(4) Where bricks are used, English bond (figure 5.36*b*) is preferred to water bond (figure 5.36*c*).

Note In water bond, the bed joints are staggered by either forming a half-course rebate around the outside of the concrete foundation, or starting the outer half-brick walling with either a course of snapped headers or a course of split bricks.

(5) Brick chambers are normally built half a brick thick where the depth to invert is less than 900 mm (figure 5.36*a*), after which one brick thick is the minimum.
(6) All pipes in inspection chambers are to be in channels discharging in the direction of the flow.
(7) A brick-on-edge arch should be formed in the brickwork over pipes more than 150 mm in

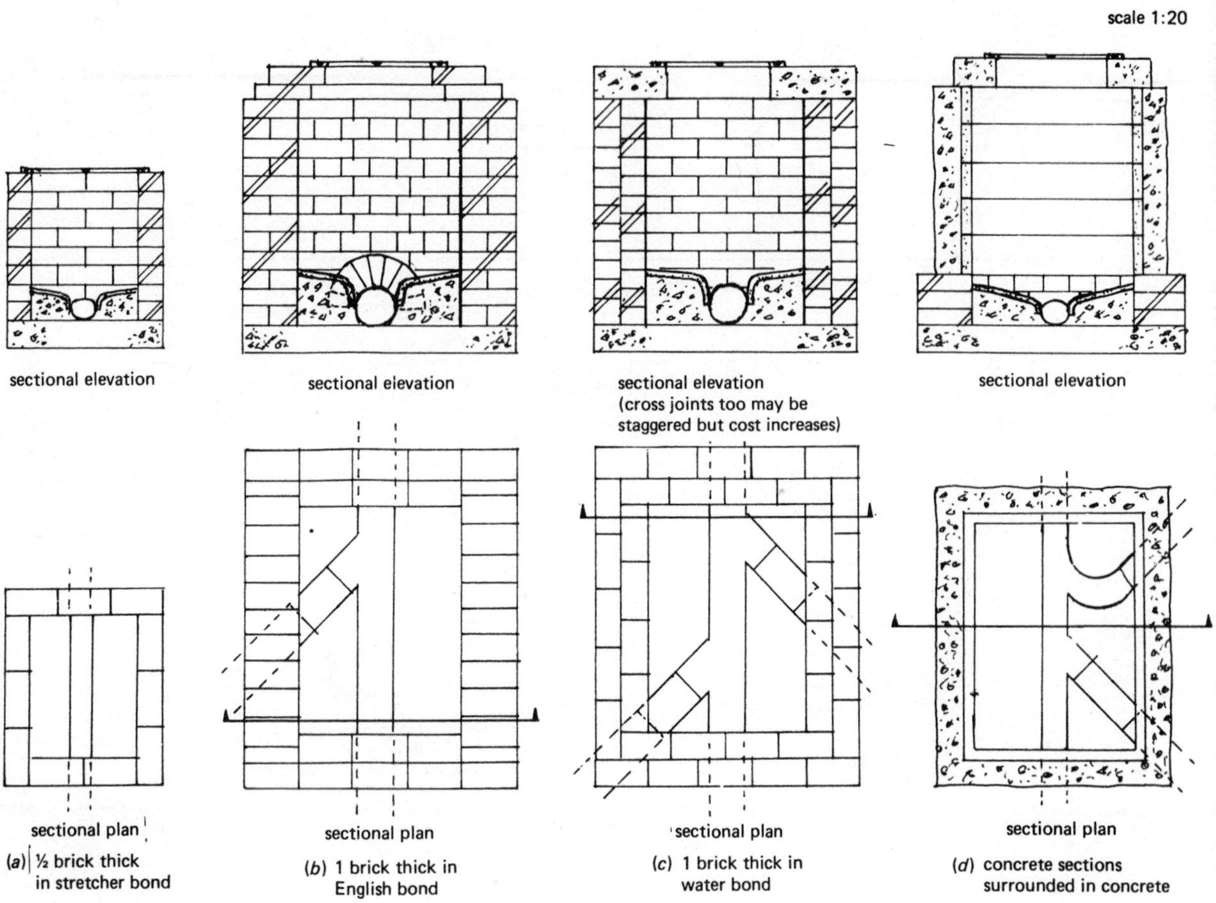

scale 1:20

sectional elevation

sectional elevation

sectional elevation
(cross joints too may be
staggered but cost increases)

sectional elevation

sectional plan

(a) ½ brick thick
in stretcher bond

sectional plan

(b) 1 brick thick in
English bond

sectional plan

(c) 1 brick thick in
water bond

sectional plan

(d) concrete sections
surrounded in concrete

Figure 5.36 Shallow inspection chambers

diameter where the chamber is deeper than 1800 mm.

(8) Benching should rise vertically on either side of the channel to the crown of the outgoing pipe, be quickly rounded off and slope upwards towards the brickwork at a slope of about 1 in 6. The mix is to be 1:1 cement and sand, trowelled smooth.

(9) The top of the chamber must be reduced as necessary to support the cover and frame, which is usually 600 x 450 mm. This is carried out by corbelling the brickwork from one or more sides (figure 5.36*b*) or with a precast reinforced concrete slab (figure 5.36*c*). It is considered good practice to install the slab below ground level to allow for completing with two courses of bricks, so that if, at a later date, the ground level is lowered the slab need not be disturbed.

(10) Where the chamber is more than 900 mm deep, step irons (figure 5.37) should be built into the walls, the vertical spacing not exceeding 300 mm

and from centre to centre 300 mm. In deep manholes a galvanised wrought-iron ladder can be used in place of step irons.

(11) At least one flexible joint on either side of an inspection chamber will help to avoid fracture in the case of ground movement.

(12) Inspection chambers are to be able to sustain imposed loads, be impervious to water and of suitable size to provide access for inspection and rodding.

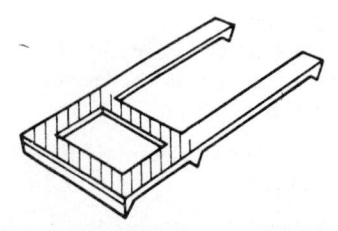

Figure 5.37 Galvanised step iron

(13) A removable, non-ventilating cover must be provided, with the frame normally bedded in mortar and the cover in grease to prevent the escape of obnoxious gases.

(14) Precast concrete sections are bedded on base sections having cut-outs for pipes, or three or four courses of brickwork are built to surround the pipework. The sections must be surrounded in at least 100 mm of concrete (figure 5.36*d*), depending on the depth.

SETTING OUT AND LAYING DRAINS

(1) Sight rails are set up behind the inspection chambers. They must be level and their height must coincide with any datum levels mentioned on the plan.

(2) Pegs are inserted to show the trench width, and the trenches are excavated by mechanical digger where the amount of work justifies their use. The last 75 mm of spoil should be got out by hand immediately before the bedding is placed. Short, shallow trenches can be excavated by hand and in each case excavating should start from the lower end, where a sump or temporary drain can be provided to prevent trenches becoming muddy in wet weather.

(3) Timbering should be carried out as necessary (see volume 1). Two further methods are shown in figures 5.38 and 5.39.

(4) Cast concrete bases for the inspection chambers; fix the channels and one pipe pointing from each channel in the correct direction.

(5) Attach a taut line and lay the pipes; the barrels must rest on the ground or in the concrete bedding and not on their collars.

(6) For rigid-jointed clay pipes, a strand of gaskin is wrapped round each spigot to centre the pipe in the collar of the previous pipe and to prevent any collaring mortar being forced through into the barrel of the pipe. A badger can be used to check this (figure 5.40). The joints should be caulked up with cement and sand in the proportion 1:2 and flaunched off at 45° (gaskin/cement joint, figure 5.18).

(7) Every fifth pipe should be boned in (figures 5.41 and 5.57); collars should be protected against the elements with sacking to prevent premature drying out.

(8) Notify the local authority before haunching or covering a drain (24 hours' notice is required).

(9) Cover the drain as required; no large stones are to be used in the first 300 mm of backfill and tamping is to be light up to this point.

(10) Send notice to the local authority not more than 7 days after the completion of backfilling.

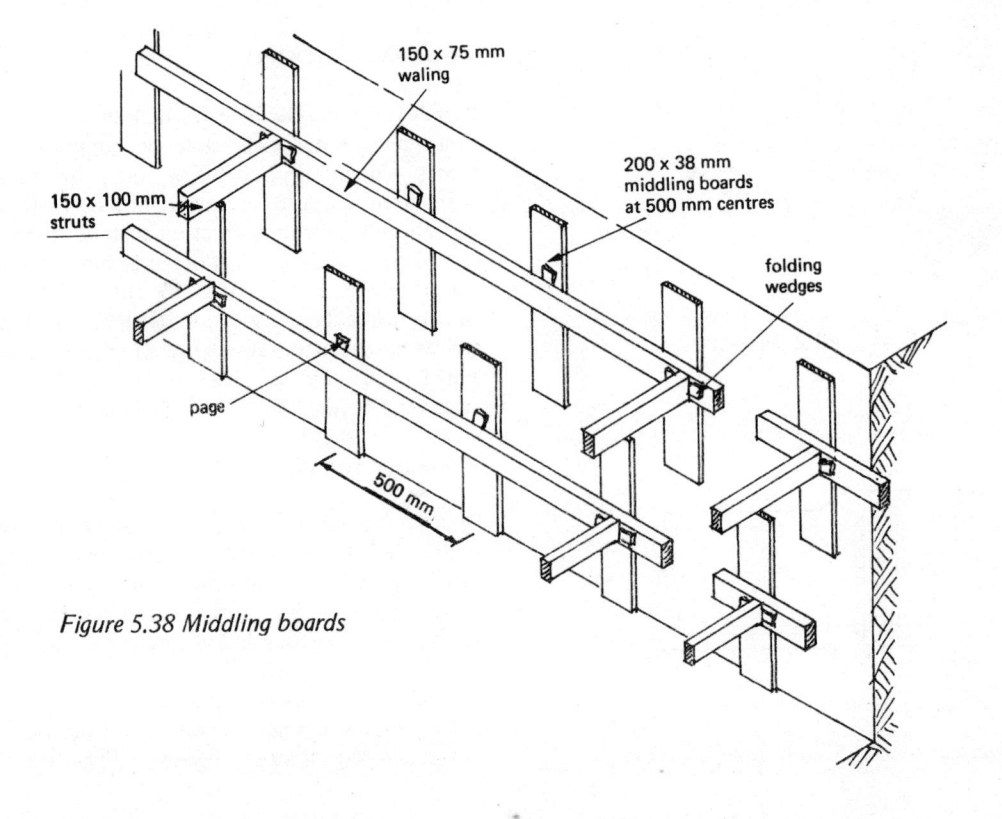

Figure 5.38 Middling boards

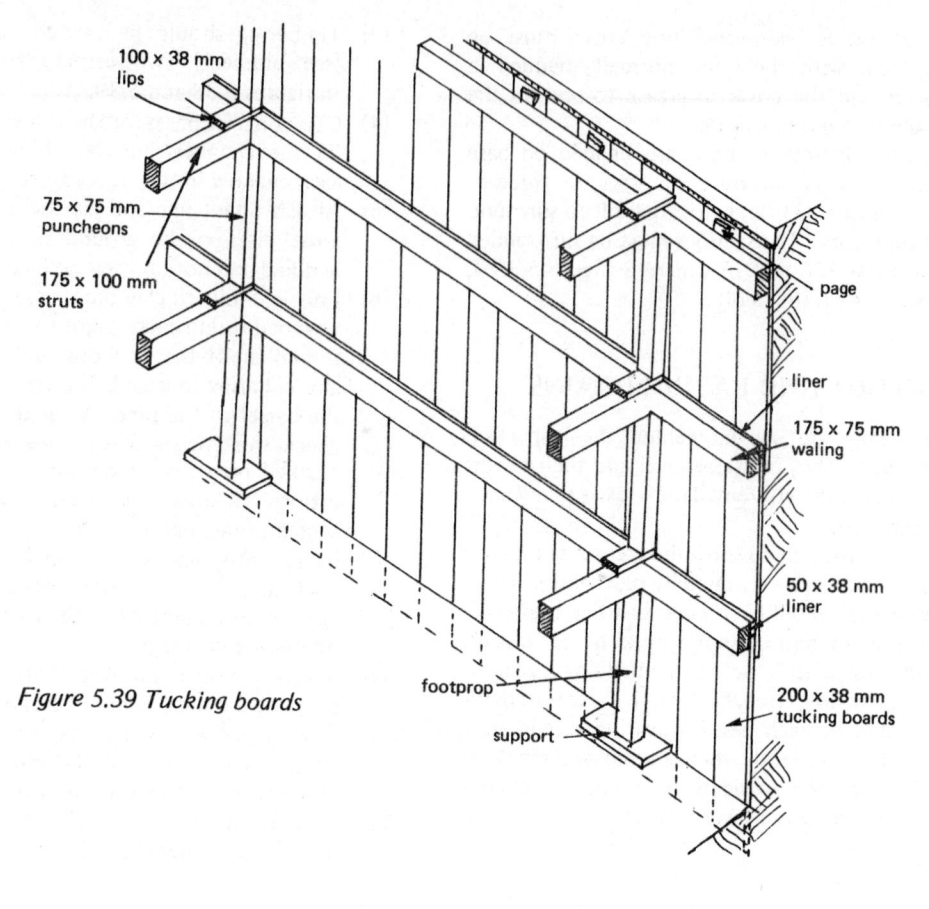

100 x 38 mm lips

75 x 75 mm puncheons

175 x 100 mm struts

page

liner

175 x 75 mm waling

50 x 38 mm liner

footprop

support

200 x 38 mm tucking boards

Figure 5.39 Tucking boards

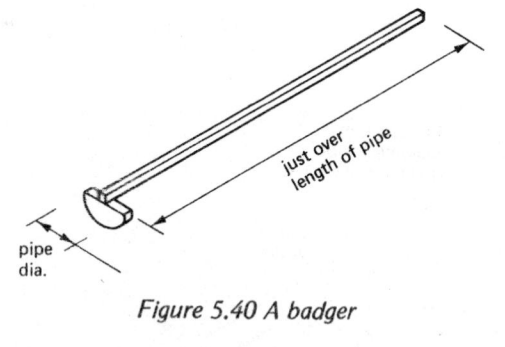

just over length of pipe

pipe dia.

Figure 5.40 A badger

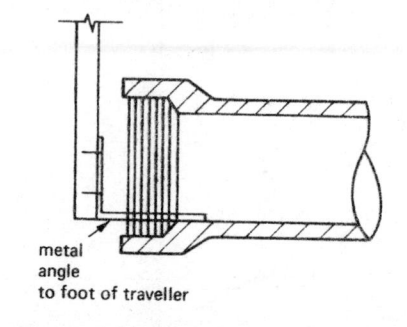

metal angle to foot of traveller

Figure 5.41 Detail of foot of traveller

TESTING DRAINS

Part N11 of the Building Regulations directs that any drain or private sewer shall on completion of the works, including backfilling, etc., be capable of withstanding a test for watertightness. Testing should be carried out from inspection chamber to inspection chamber, including any short branches; long branches should be tested separately. The length of drain between the last inspection chamber on site and a saddle on the main sewer should be tested via a testing junction installed close to the main sewer, which is sealed off before backfilling takes place.

The Water Test (figure 5.42)

This is the most widely used test and is generally considered to be the most reliable. Where rigid joints have been used it is important that at least 24 hours should elapse before testing, to allow the mortar to gain sufficient strength, but flexibly jointed pipes can be tested immediately.

The test is applied by plugging the lower end of the pipeline with either an expanding rubber ring plug or an air bag stopper (figures 5.43 and 5.44), and

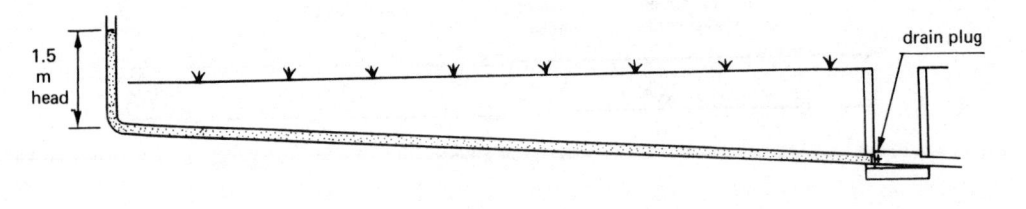

Figure 5.42

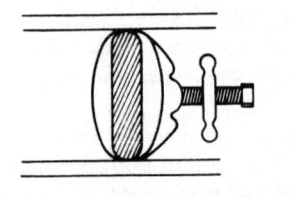

Figure 5.43 Expanding drain plug

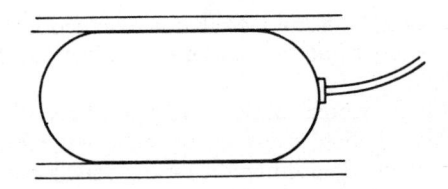

Figure 5.44 Inflated air-bag stopper

filling the pipeline with water to provide a head of 1.2 m above the higher end. It is important not to subject the lower end to more than a 6 m head to avoid overstressing the joints. After the initial loss of head, due to absorption by pipes and joints, the movement of trapped air and sweating at the joints, the head should be held steady with no further loss for at least 10 minutes to pass the test.

Note   Where this test is being carried out in water-logged trenches, colouring powder should be placed in the testing water, and any leaks will then be quickly noticed.

The Air Test (figures 5.45 and 5.46)

Where water is not available or its disposal is inconvenient, this test is considered to be a good alternative.

The test is carried out by firmly plugging each end of the pipeline and pumping in air until a pressure of 100 mm is indicated on the manometer. The pressure should not fall from 100 mm to below 75 mm during a period of 5 minutes. If the pressure does fall below this a leak is indicated, and if this cannot be located, a smoke bomb can be used.

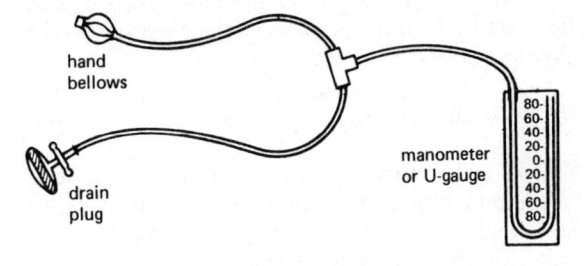

Figure 5.46

The Smoke Test

A smoke bomb or smoke-generating machine is used to supply smoke at the lower end of the system. The top of the vent pipe is plugged and the seals in the gullies are removed until smoke is seen emerging. The water seals are replaced and the seal at the top of the vent pipe is removed until smoke emerges. The seal is

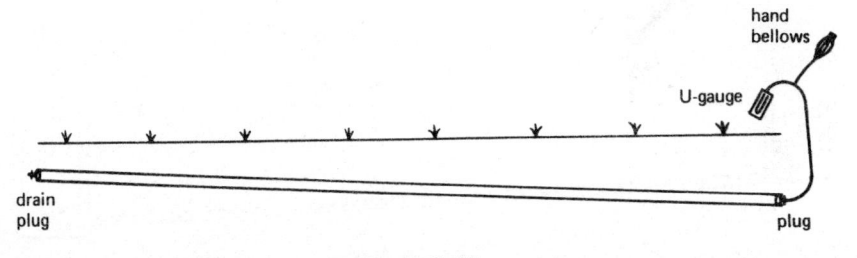

Figure 5.45

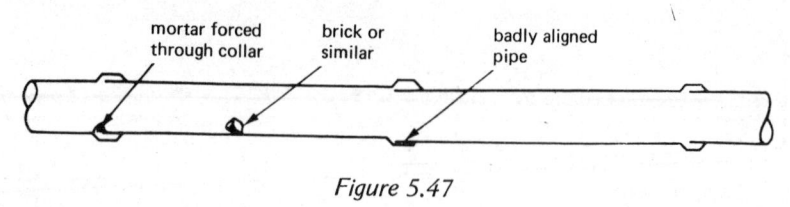

Figure 5.47

reinstated and smoke continues to enter the pipeline. This test is considered to be imprecise and is not recommended.

While the three tests mentioned will show any leaks that exist, there may be internal obstructions within the pipeline that will cause a blockage in the course of time (figure 5.47). This can be checked in two ways

The Ball Test

A smooth, solid rubber ball, 13 mm less in diameter than the bore of the pipe, is inserted into the top end and should roll freely down the invert of the pipe. If it stops, a blockage is indicated. This is located by inserting drain rods into the pipeline until they touch the ball; the rods are then removed and laid alongside the drain to show the position of the blockage, which should be corrected as necessary.

The Reflection Test

A lamp and a mirror should be placed in the inverts of adjacent inspection chambers, as shown in figure 5.48. Light is reflected along the drain and the condition of the bore can be examined. This test can only be used where the drain is perfectly straight.

REPAIRS AND ALTERATIONS TO DRAINS

Inserting a New Pipe or Junction in a Straight Run

Occasionally a pipe may be damaged, possibly as a result of excavations, or a new set of pipes has to be connected into an existing run. There are two usual methods of carrying out this work.

Method 1

(1) Expose the broken pipe and one more pipe on either side of it (figure 5.49*a*).
(2) Break out the three pipes, taking great care not to damage the collar and spigot on either side (figure 5.49*b*).
(3) Insert the new pipes as shown in figure 5.49*c*. They will drop into position and can be lined up and jointed (figure 5.49*d*).

Method 2

(1) Expose both the pipe that is to be replaced with a junction and the pipe above it in the run (figure 5.50*a*).
(2) Carefully break out the two pipes (figure 5.50*b*).
(3) Slide a loose collar on to the spigot remaining and place in position the junction and a butt pipe (figure 5.50*c*).
(4) Slide the loose collar over the joint, line up and make good all joints (figure 5.50*d*).

Connections to Inspection Chambers

(1) Cut away the benching on the appropriate side.
(2) Cut a suitably sized hole in the side wall.
(3) Bed the required splayed-end channel bend in position and connect the new pipe run to this.
(4) Make good to the walls and re-form the benching.

Connecting to a Main Sewer using a Saddle

(1) Check the depth of flow in the main at a convenient inspection chamber and cut a hole in the cheek above the flow to avoid surcharging the branch (figure 5.51). This can be carried out with a small hammer and a sharp chisel.

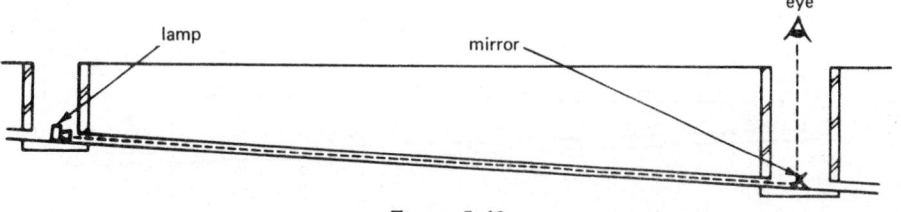

Figure 5.48

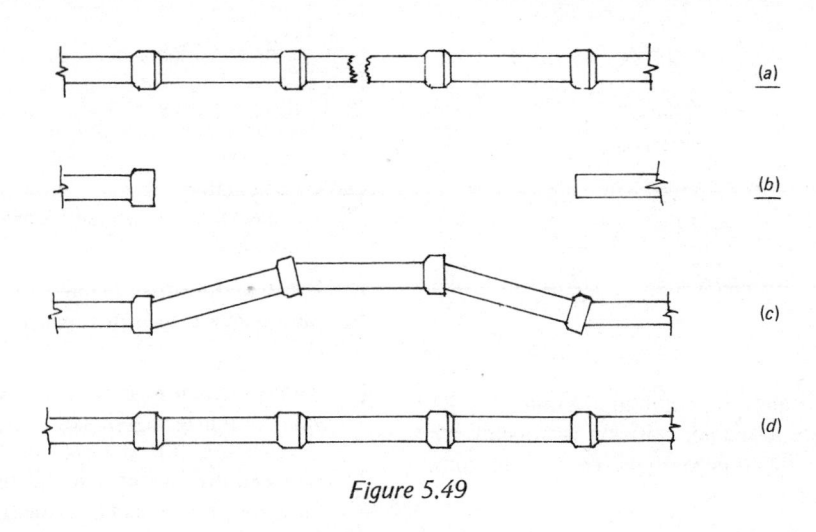

Figure 5.49

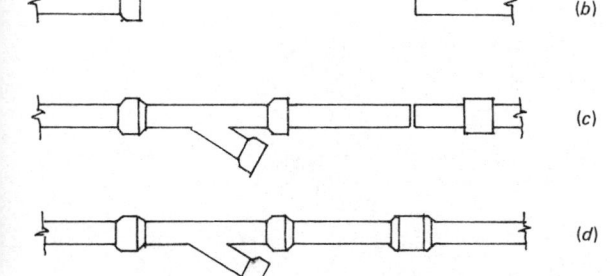

Figure 5.50

Note The local authority will normally inspect the connection before concreting is carried out and some authorities require the insertion of a testing junction close to the sewer connection in order to test the complete system. After testing, the junction is sealed off.

Cutting Drainpipes

Apart from the use of a special cutting tool, the following two methods are in common use.

(1) Stand the pipe on end on a reasonably soft surface and fill with sand up to the required cutting mark (figure 5.52). Tap round the pipe on the cutting mark with a small hammer and sharp chisel, or a scutch, until fracture occurs.

(2) Carefully enlarge the hole with the same tools and/or a scutch, cutting at the thickness of the main rather than on the face.

(3) When the spigot of the saddle fits snugly, clean any pipe debris out of the main and bed the saddle in cement mortar.

(4) Putting your hand through the saddle, clear away any mortar that has squeezed through, connect the pipework and surround it in concrete.

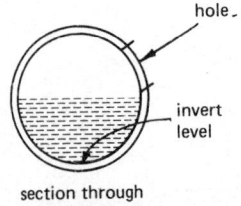

hole

invert level

section through main sewer

Figure 5.51

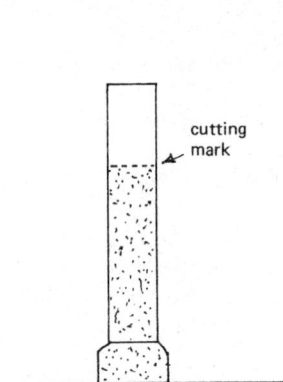

cutting mark

Figure 5.52

(2) Form a mound of sand and with the barrel of the pipe resting on it at the cutting mark and with the spigot unsupported, tap as before while rotating the pipe (figure 5.53).

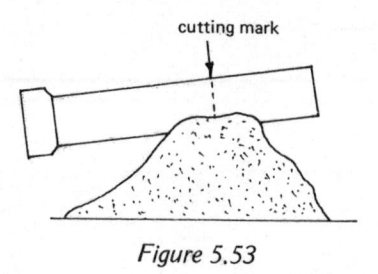

Figure 5.53

Note   Where channels are required but only straight pipes or bends are available, splitting is possibly by carefully tapping the pipe along each side in turn until fracture occurs.

CALCULATION OF INVERT LEVELS

Before any pipes are laid it is necessary to know the invert level of the drain or sewer to which the connection is to be made and to relate this to the ground level point at which pipelaying is to start. The invert level is arrived at in one of three ways

(1) the local authority may be able to provide the information
(2) by calculation
(3) by excavation vertically downwards to expose the main

If the local authority is unable to provide the information and excavation is inconvenient, calculation is as follows.

(1) Remove the inspection chamber covers on either side of the proposed saddle position and with a Cowley or tilting level obtain the difference between the invert levels (figure 5.54).
(2) Measure the overall distance between these inspection chambers and the distance from each to the saddle (figure 5.55).
(3) Calculate the invert level of the drain at the saddle as follows. Referring to figures 5.54 and 5.55, since

$$\text{total fall} = 1 \text{ m } (2.6 - 1.6)$$

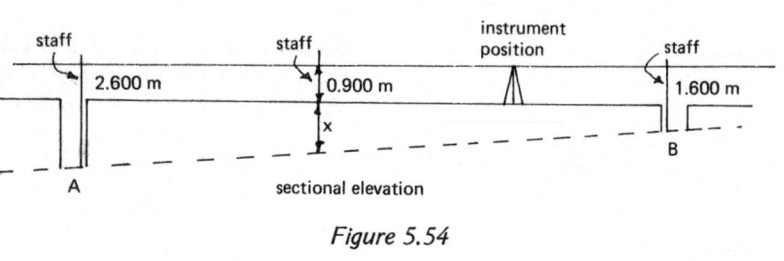

Figure 5.54

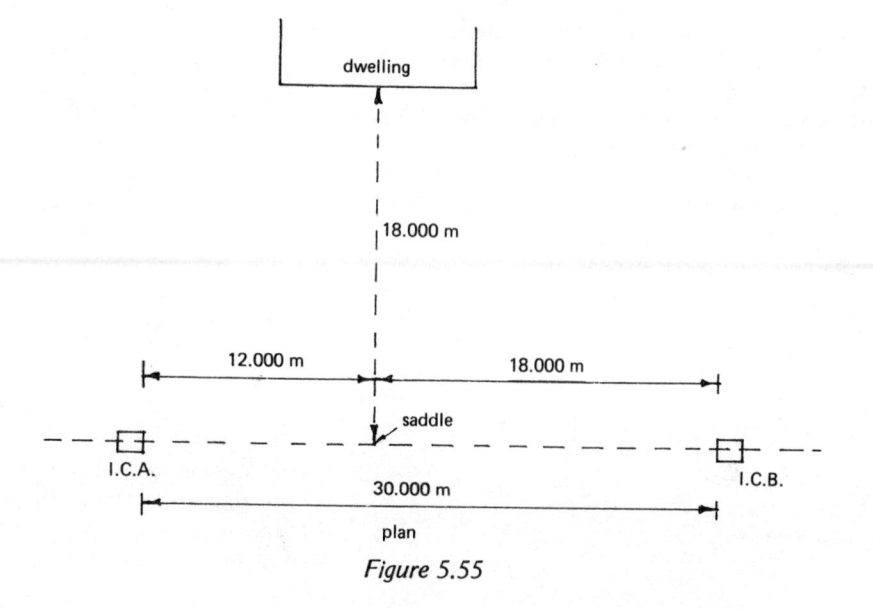

Figure 5.55

the main is laid at a fall of 1 in 30 (the distance between inspection chambers is 30 m). To find the fall in 18 m

$$\frac{1}{30} = \frac{fall}{18}$$

Cross multiplying

$$30 \times fall = 1 \times 18$$

dividing both sides by 30

$$\frac{30 \times fall}{30} = \frac{1 \times 18}{30}$$

Cancelling

$$fall = \frac{3}{5} = 0.6 \text{ m}$$

Thus the fall from inspection chamber B to the saddle is 0.6 m.

Note It may be simpler at this stage for the student to remember that

$$fall = \frac{\text{actual distance}}{\text{given distance}} = \frac{18}{30} = 0.6 \text{ m}$$

Since

staff reading at B = 1.6 m

depth of invert at saddle = 1.6 + 0.6 (depth at B + fall)

$$= 2.2 \text{ m}$$

And since

ground level staff reading at the saddle = 0.9

depth of invert below ground level = 2.2 − 0.9

$$= 1.3 \text{ m}$$

That is

$$x = 1.3 \text{ m}$$

Having found this it is necessary to relate it to the invert level at the topmost inspection chamber of the proposed drain to ascertain that the fall will be suitable. Assume that the invert level is 0.6 m below ground level (see figure 5.56). An instrument is set up between these points and readings are taken of 1.600 and 1.4 as shown. Since the invert level at X is 0.6 below ground level

fall from invert level at X to invert Y

$$= 0.5 \text{ m} (2.7 - 2.2)$$

Thus the fall is 1 in 36, which is quite satisfactory (0.5 in 18 m equals 1 in 36). Sight rails can now be set up at each end (figure 5.57), and the

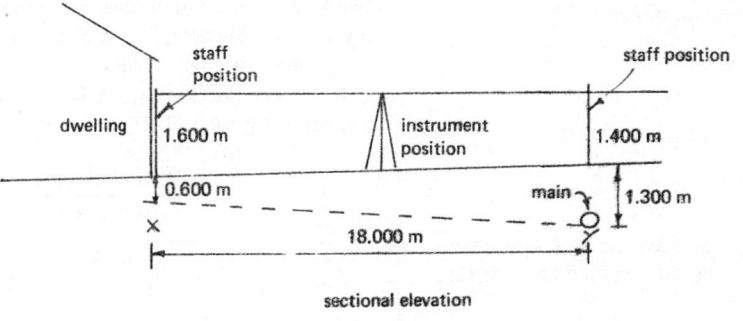

sectional elevation

Figure 5.56

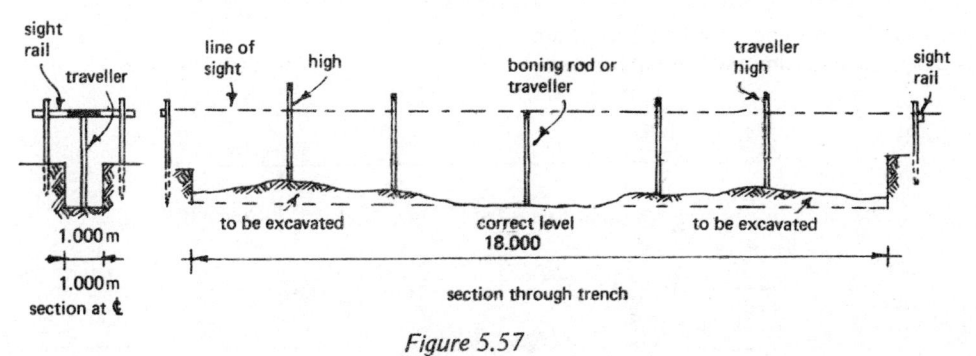

section through trench

Figure 5.57

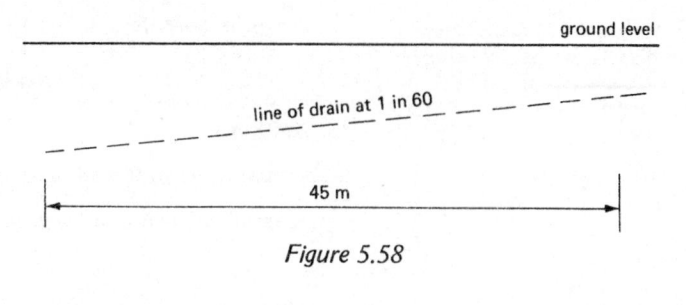

Figure 5.58

ground is excavated. Boning rods are used to obtain the required slope and the pipes are laid.

Note   If, when the depth of the inspection chambers is checked in the first place, it is obvious that there will be an adequate fall, the above procedure may be unnecessary.

While the foregoing explanations on invert levels were necessary, it is probable that all that the craft student will be required to do is to calculate the total fall between inspection chambers, given the overall length and required fall.

Example 5.4

A drain is to be laid a distance of 45 m in level ground and the fall is to be at 1 in 60. Calculate the total fall (figure 5.58).

$$\text{Total fall} = \frac{\text{actual distance}}{\text{given distance}}$$

$$= \frac{45}{60}$$

$$= 0.75 \text{ m}$$

$$= 750 \text{ mm}$$

Possibly the invert level may have to be found given the level at one end, the overall distance and the fall.

Example 5.5

The invert level at the top end of a length of drain 50 m long is 2.50 above datum and the fall has to be

1 in 40. Find the invert level of the bottom inspection chamber (figure 5.59).

$$\text{Total fall} = \frac{\text{actual distance}}{\text{given distance}}$$

$$= \frac{50}{40}$$

$$= 1.25 \text{ m}$$

Since

invert level at top end = 2.50

then

invert level at bottom = 2.50 − 1.25

$$= 1.25 \text{ A.D. (above datum)}$$

Use of a Tapered Straightedge

Another method of obtaining the correct fall is to use a tapered straightedge in conjunction with a spirit level. For example, if a fall of 1 in 40 is required, a 4 m straightedge is ideal. This should be about 150 x 25 in section and it is cut down lengthways as shown in figure 5.60. That is

$$\text{fall} = \frac{\text{actual distance}}{\text{given distance}}$$

$$= \frac{4 \text{ (4 m straightedge)}}{40 \quad \text{(1 in 40)}}$$

$$= \frac{1}{10} \text{ m}$$

$$= 100 \text{ mm}$$

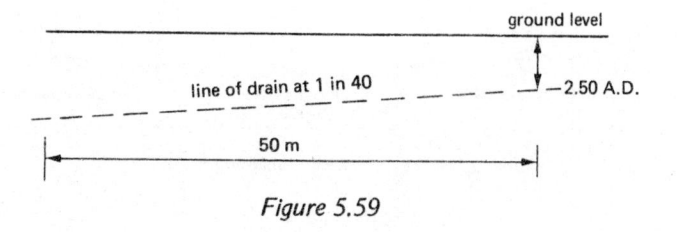

Figure 5.59

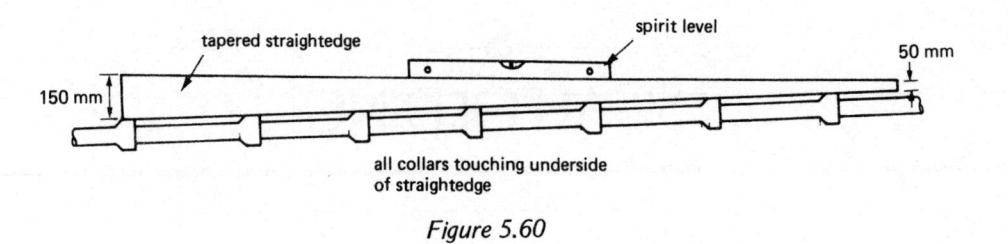

Figure 5.60

Therefore, the straightedge should taper by 100 mm, say from 150 to 50 mm, in 4 m, and a spirit level should be placed on top centre to check the fall (figure 5.60).

6
SCAFFOLDING

CONSTRUCTION REGULATIONS

Where men are unable to reach their work from the ground or part of a building, ladders or a scaffold must be provided. Tubular scaffolding is a temporary structure, erected to support a platform or number of platforms, at different heights, and may be of steel or aluminium alloy. Its erection is strictly governed by the Construction (Working Places) Regulations, which came into operation in August 1966 and have not as yet been metricated; therefore any measurements given have been converted.

It is the duty of every employer and employee to comply with the requirements of the Construction Regulations, which are summarised as follows.

(1) A sufficient quantity of materials is to be provided, to be sound and of adequate strength for its purpose.
(2) The erection, alteration and dismantling of a scaffold must be carried out under the supervision of a competent person.
(3) All materials intended for a scaffold must be inspected by a competent person, who must also inspect the completed scaffold at least every 7 days and after exposure to adverse weather conditions. The results of an inspection must be entered in the prescribed register (figure 6.1).
(4) Any timber to be used must be in good condition, of suitable quality and not painted in such a way that defects are hidden.
(5) Scaffolds must not be overloaded and materials are not to be kept on a scaffold unless required within a reasonable time.
(6) Partly dismantled scaffolds must comply with the Construction Regulations or carry permanent warning notices. The access to incomplete scaffolds should be effectively blocked as far as possible.
(7) Loose materials such as bricks, drainpipes, chimney pots, etc. must not be used as supports for platforms, but a firm packing of bricks or blocks may be used if stable up to a height of 600 mm above ground level.
(8) A platform must extend beyond the end of a

SCHEDULE   Regulation 22

FACTORIES ACT 1961

CONSTRUCTION (WORKING PLACES) REGULATIONS 1966

SCAFFOLD INSPECTIONS

FORM OF REPORTS OF RESULTS OF INSPECTIONS UNDER REGULATION 22 OF SCAFFOLDS, INCLUDING BOATSWAIN'S CHAIRS, CAGES, SKIPS AND SIMILAR PLANT OR EQUIPMENT (AND PLANT OR EQUIPMENT USED FOR THE PURPOSES THEREOF)

Name or title of Employer or Contractor ....

Address of Site ....

Work Commenced—Date ....

Location and Description of Scaffold, etc. and other Plant or Equipment Inspected (1)	Date of Inspection (2)	Result of Inspection. State whether in good order (3)	Signature (or, in case where signature is not legally required, name) of person who made the inspection (4)

Figure 6.1

wall at least 600 mm if work is to be carried out at that point.

TUBES, FITTINGS AND BOARDS

Members (Tubes)

Standards

Standards are the upright members of a scaffold and they are usually spaced between 1.8 and 2.5 m apart, depending on the load to be carried and the type of work being done. They must be vertical or slightly inclined towards the building and sufficiently close to ensure stability. A firm base is essential and they can be extended where required, using joint pins or sleeve couplers, the height of which should be staggered.

Ledgers

Ledgers are long, horizontal members, which are fastened to the standards on the inside using load-bearing couplers. They are normally secured together lengthways with sleeve couplers or joint pins, and if the latter are used they must be positioned at one-third of the bay owing to their lack of tensile strength.

Putlogs

Putlogs are short, flat-ended tubes 1.2 to 1.5 m long, which are inserted into the bed joints of brickwork to the full extent of the flat supporting surface. They are usually fastened to the ledgers with putlog couplers and in each bay one putlog must be within 300 mm of a standard. The spacing of the putlogs varies according to the thickness of the planks

Plank thickness (mm)	32	38	50
Putlog spacing (m)	1	1.5	2.5

Transoms

Transoms are short lengths of tube that take the place of putlogs in an independent scaffold, both ends being supported by ledgers to which they are secured using putlog couplers.

Longitudinal Braces

Longitudinal braces are lengths of tube fastened at or as near to 45° as possible on the outside of standards to provide stability and eliminate sideways movement. They are required every 30 m and must extend to the full height of the scaffold. It is preferable to fasten them to putlogs or transoms with double couplers, or alternatively to standards with swivel couplers.

Cross Braces

Cross braces are short lengths of tube used to connect and give added rigidity to alternate pairs of standards in an independent scaffold. They can be fastened to ledgers with double couplers or to standards with swivel couplers.

Puncheon

A puncheon is a short length of vertical tube that does not touch the ground.

Bridle Tube

A bridle tube is a horizontal tube secured just clear of the wall face in a putlog scaffold. It is secured across openings below the putlogs on either side with double couplers and is used to support extra transoms as required to carry the boards (figure 6.2). Where a wide opening occurs the centre transom can be supported off the window bottom if necessary.

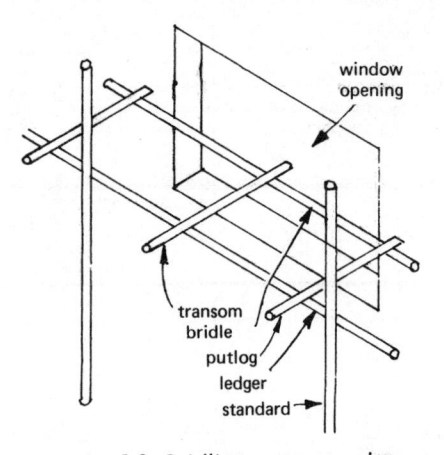

Figure 6.2 Bridling on an opening

Guard Rails (figure 6.3)

Guard rails are lengths of tube which must be provided where men are liable to fall more than 2 m. They must be secured on the inside of the standards at a height of between 0.9 m and 1.125 m and kept permanently in position except for access and loading.

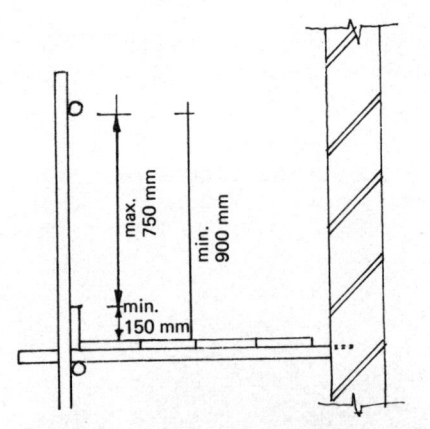

Figure 6.3 Guard rails and toe boards

Raking Tube or Raker

This is a length of tube which can be used to prop up a scaffold initially before the insertion of reveal or through ties. Where a wall contains no openings for ties, rakers provide an alternative method of preventing a scaffold from pulling away from the wall. They should be fixed at 45° or as near as possible using double or swivel couplers and be provided with a sound foot block at the base (figure 6.24).

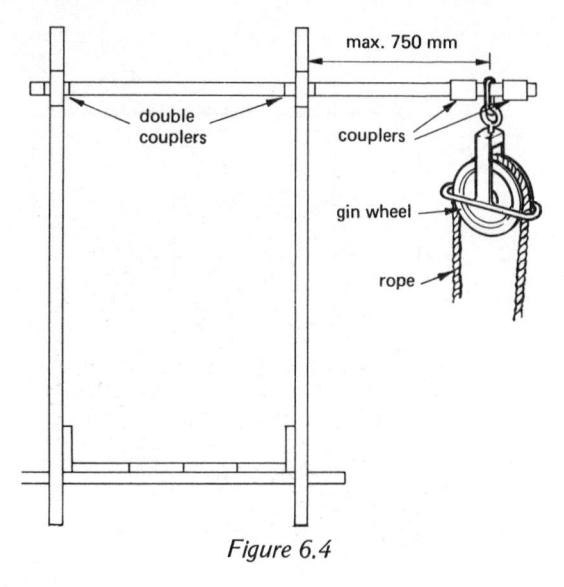

Figure 6.4

Butting Pieces

These are very short lengths of tube used, for example, to reinforce across a joint pin. A butting piece should be securely fastened on either side with parallel or universal couplers.

Scaffold Fittings

Many types of fitting are produced by different firms; those shown in figures 6.5–6.20 are S.G.B. scaffold fittings.

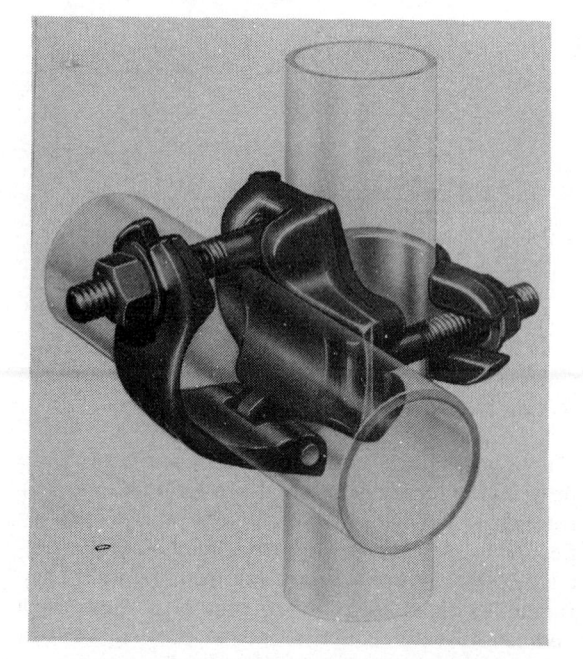

Figure 6.5 Drop-forged double coupler

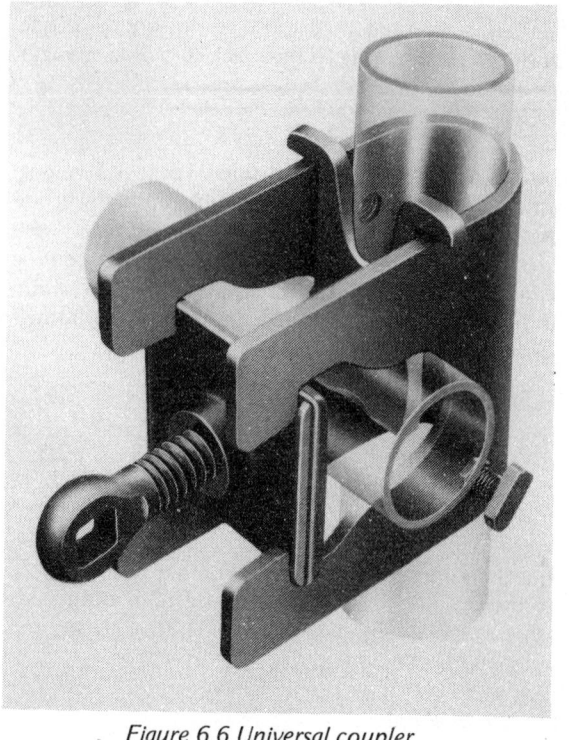

Figure 6.6 Universal coupler

Double Couplers (figure 6.5)

These are used for fastening ledgers to standards and in all positions where strength is required, for example, bridle tubes to putlogs.

Universal Coupler (figure 6.6)

This is another 90° coupler, which can also be used for connecting two loadbearing tubes in parallel.

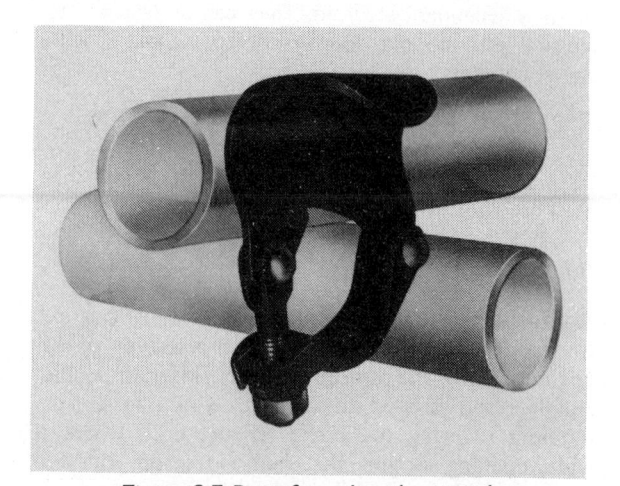

Figure 6.7 Drop-forged putlog coupler

Figure 6.8 Putlog or brace coupler

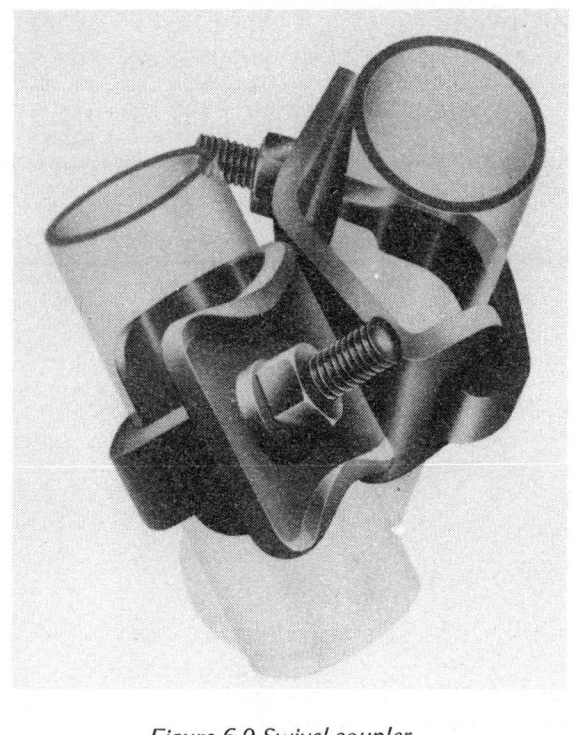

Figure 6.9 Swivel coupler

Putlog Couplers (figures 6.7 and 6.8)

Used for connecting putlogs or transoms to ledgers.

Swivel Coupler (figure 6.9)

This is a one-piece coupler used for connecting two scaffold tubes at any angle through 360°, for example, longitudinal or cross braces to standards.

Putlog End (figure 6.10)

This is a simple fitting which will convert a transom into a putlog. When a putlog scaffold is being erected against an existing brick building it is easier to cut the putlog holes where required, insert the putlog ends secured with hardwood wedges and fit transoms to the putlog ends as the scaffold is erected.

Sleeve Coupler (figure 6.11)

This is an external fitting used to join standards, ledgers, longitudinal braces and guard rails end to end.

Joint Pin (figure 6.12)

The uses of a joint pin are as for a sleeve coupler, but it is not used for braces. It fits internally into the end of a scaffold tube and expands against the wall of the

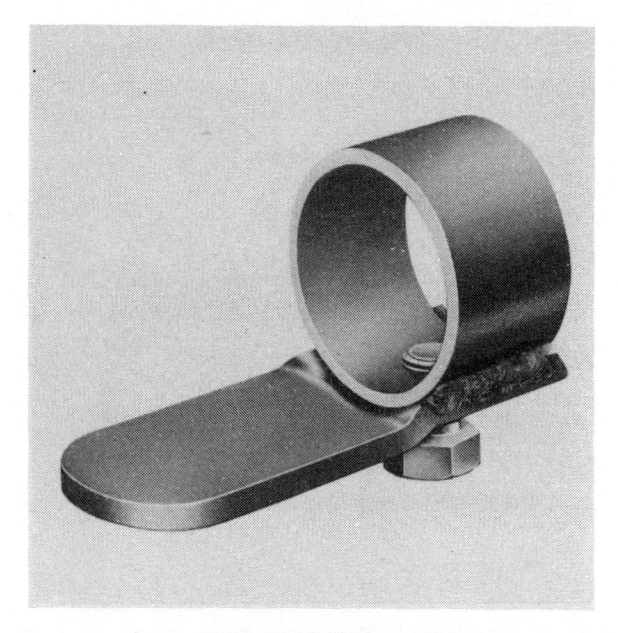

Figure 6.10 Putlog end

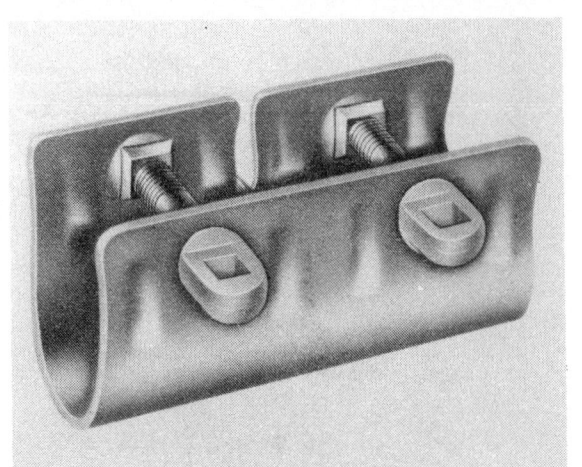

Figure 6.11 Sleeve coupler

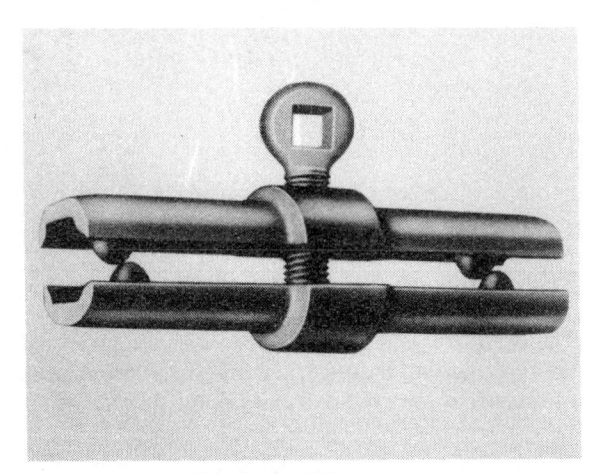

Figure 6.12 Joint pin

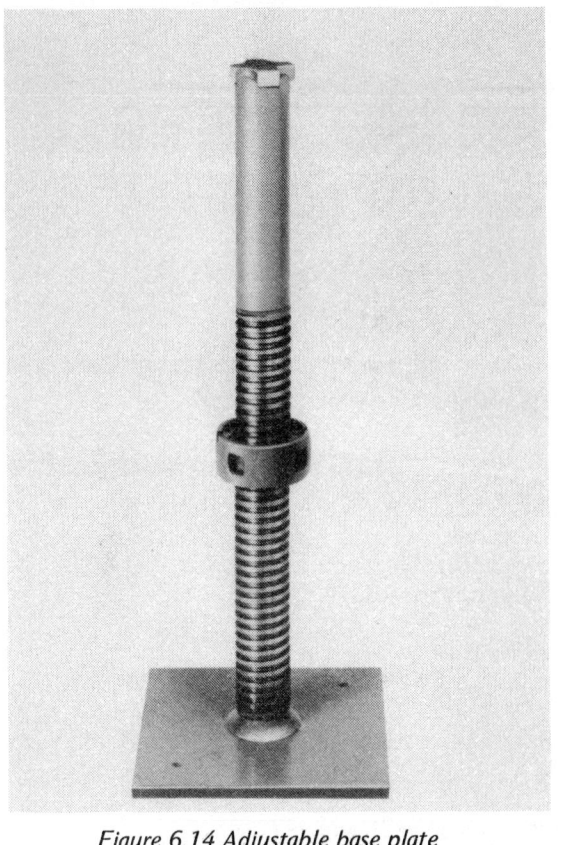

Figure 6.14 Adjustable base plate

tube as the bolt is turned. They are not as strong as sleeves and must be used at one-third the bay width, never at mid-bay. If a joint pin should occur at mid-bay it must be reinforced with a butting piece, secured on either side with a universal coupler.

Figure 6.13 Base plate

Figure 6.15 Reveal pin

Base Plate (figure 6.13)

This is a 150 x 150 mm steel plate which is used to provide a flat, bearing surface for load distribution from standards. It has a central spigot 50 mm high, on which the foot of the standard is located, and two fixing holes for use with sole plates.

Adjustable Base Plate (figure 6.14)

This is for use in undulating ground, particularly where settlement may take place. It has 230 mm of adjustment.

Reveal Pin (figure 6.15)

This fitting is inserted into the end of a short tube and is expanded as necessary to form a rigid horizontal or vertical tie in a window opening to which the scaffolding can be secured.

Fixed Finial (figure 6.16)

This is used to connect a scaffold tube at right-angles to the extreme end of another tube without projection. It is very useful for guard rails, safety barriers, etc.

Toeboard Clip (figure 6.17)

This is used to secure a toeboard against a standard.

Figure 6.17 Toeboard clip

Gin Wheel (figure 6.18)

The figure shows a 250 mm steel wheel, with which a 19 mm-diameter rope is used. Its safe working load is 250 kg.

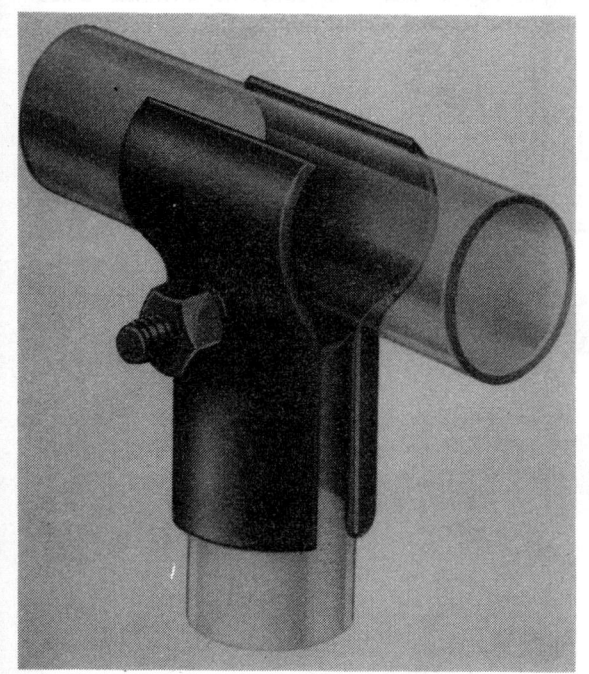

Figure 6.16 Fixed finial

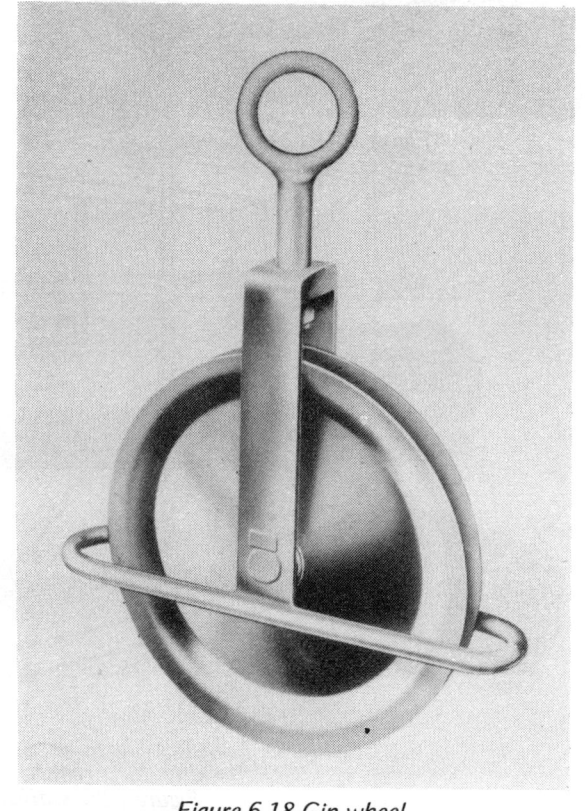

Figure 6.18 Gin wheel

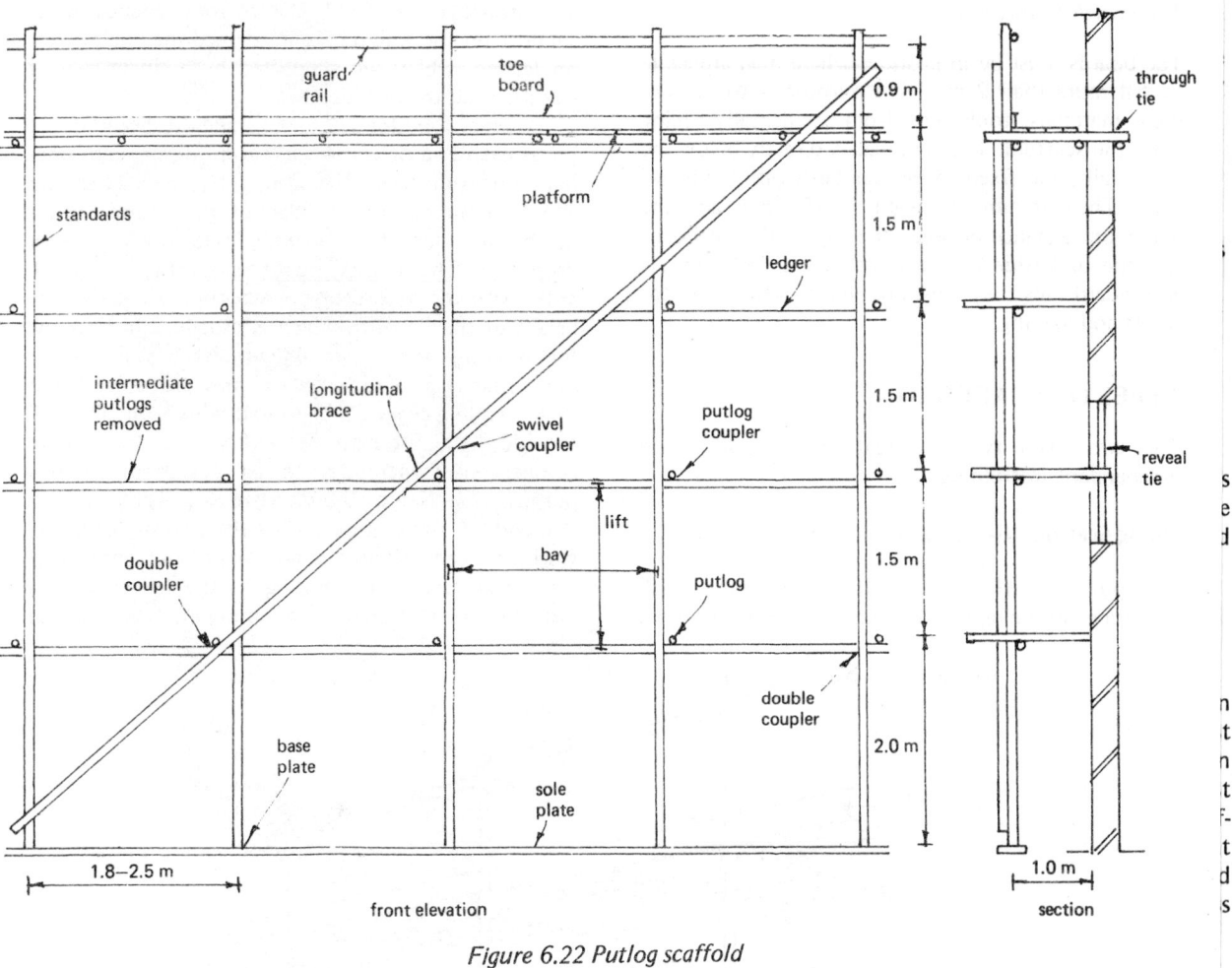

Figure 6.22 Putlog scaffold

on the platform as close to the standard as possible the stress in the ledgers will be reduced to a minimum.

Fix intermediate putlogs as required with due regard to the thickness of the planks, and plumb and level the scaffolding as it is erected, tightening fittings as work proceeds. When the length of the standards exceeds 6.5 m joint pins or sleeve couplers are used to connect the tubes together. As already mentioned, it is important not to have all the joints occurring at the same height, but to stagger them by using tubes of different lengths. With a putlog scaffold only one lift must be in use at a time.

Independent Scaffold (figures 6.23 and 6.24)

The independent scaffold is normally used on existing buildings or on structures where putlogs would be inconvenient. It is so called because it is self-support-ing and carries all the superimposed loads without

assistance from the structure. It consists basically of two rows of standards, two rows of ledgers, transoms, longitudinal braces and cross braces.

When erecting this scaffold for bricklayers, set the inner row of standards about 330 mm from the wall so that the inside scaffold board can be placed on the transoms, projecting beyond the inner standards. Lifts are usually approximately 2 m. It is possible for one man to set up a simple independent scaffold using the 'mattress' method of erection as follows.

(1) Construct a temporary base frame, the length and width of the required scaffold, and pack up level on bricks, blocks, etc. about 600 mm above the ground.
(2) Fix the four end standards to this frame with double couplers.
(3) Fasten ledgers to these standards and transoms to the ledgers where required, plumbing and levelling as work proceeds.

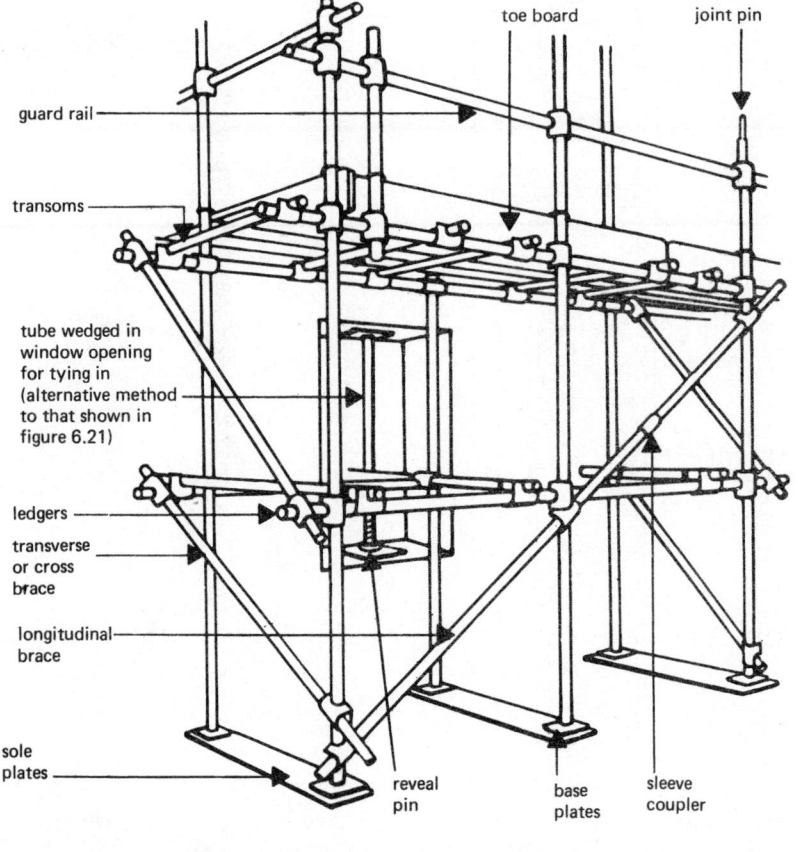

Figure 6.23 Independent tied scaffold

(4) Intermediate standards are now positioned, taking any sag out of the ledgers as necessary.
(5) Longitudinal and cross braces can now be fixed and the base frame removed.

Note Both types of scaffold must be securely tied to the building at least every 4 m vertically and 6 m horizontally. This is usually carried out in one of two ways

(1) with transoms passing through window or other openings, connected to tubes fastened with double couplers at right-angles to these inside the structure and close up against the wall (figure 6.21); or
(2) with short lengths of tube wedged in window openings with reveal pins; not more than 50 per cent of the ties may be of this type (figure 6.23).

Where openings are non-existent and the height of the scaffold is limited, the scaffold should be strutted from the ground with raking tubes inclined towards the building (figure 6.24).

Widths of Working Platforms

Where an operative is liable to fall more than 2 m, the Construction (Working Places) Regulations lay down minimum platform widths of not less than

(1) 625 mm (loosely described as three boards wide) where the platform is used as a footing only, that is, not for depositing materials
(2) 850 mm (loosely described as four boards wide) where used for working from, and for depositing materials
(3) 1 m where used to support a higher platform
(4) 1.3 m where used for dressing stone
(5) 1.5 m where used to support a higher platform, and for dressing stone.

Responsibility for Scaffolds

Where a scaffold is erected by one employer and used by, or on behalf of, another employer, it is the responsibility of the first mentioned to ascertain that the scaffold and the materials from which it is con-

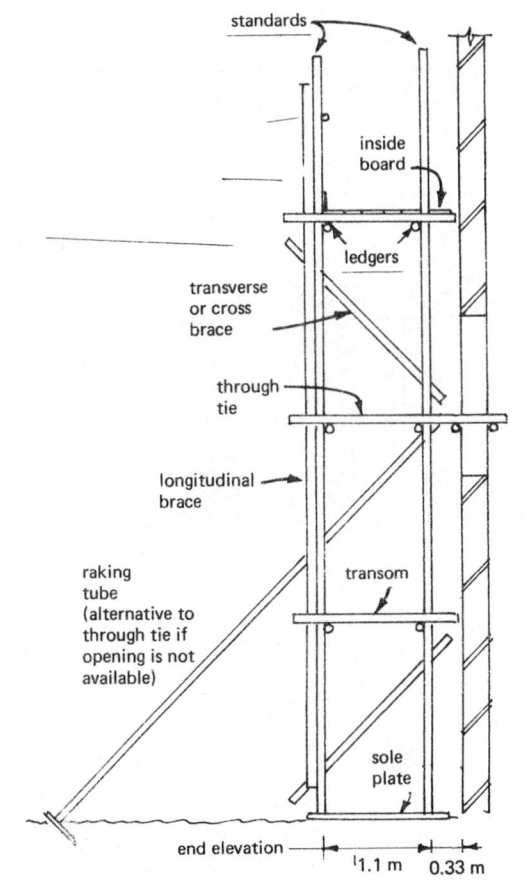

Figure 6.24 End elevation of independent scaffold

structed are sound and stable and that the Construction Regulations are kept to throughout.

Treatment and Storage of Equipment

Tubes should be stacked in racks in lengths, clear of the ground, and protected against the weather. Fittings should be cleaned and lightly oiled and stored in separate bins. When a scaffold is being dismantled, tubes should be carefully lowered to the ground since bent tubes may not be re-used and fittings require at least as much care to prevent loss or damage.

Trestle Scaffolds

Many types of adjustable steel trestle are available. Those shown in figure 6.25 are by S.G.B. Ltd and are designed in three widths to take three, four or five boards. The advantages of using trestle scaffolds include the following.

(1) They are light in weight, but strong.

Figure 6.25 S.G.B. adjustable steel trestles

(2) They are easily and quickly set up by one man.
(3) The widely splayed feet give stability, the legs being immovable when in use because of the fixed locating lugs.
(4) Adjustment of height is simple and positive and each of the four sizes available will extend to at least three-quarters of the initial height.
(5) The rests for the boards are flat and the pins are securely attached to the frames and cannot be lost or misplaced.
(6) Storage is facilitated by turning the splayed feet through 90°.
(7) They are very useful for single internal lifts in housing.

The Construction Regulations appertaining to trestle scaffolds lay down a maximum height of 4.5 m, which is generally considered to be three lifts; the trestles should be adequately braced to prevent sideways movement. Figure 6.26 shows a trestle scaffold

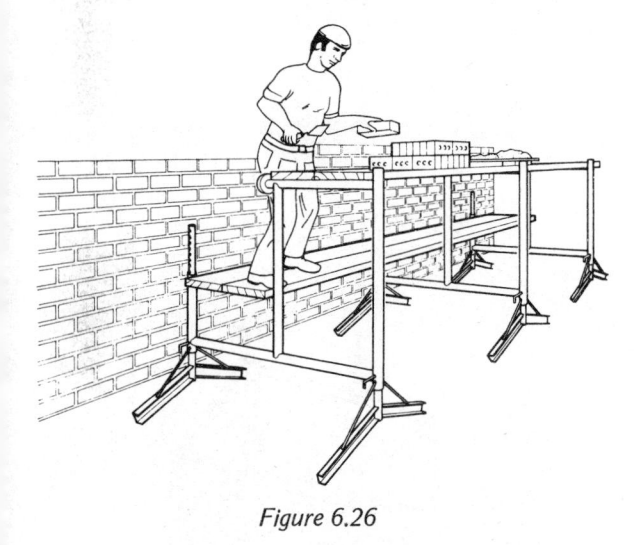

Figure 6.26

with a raised platform for depositing materials, which eliminates a lot of the bending normally associated with bricklaying.

Tower Scaffolds

An independent tower scaffold, apart from the necessary ties, stands completely free from buildings and is mostly used for overhead maintenance work where only a small working area is required.

It consists basically of standards, ledgers, transoms, diagonal bracing and plan bracing. Some of the requirements for independent tower scaffolds are as follows.

(1) Standards should be on base plates with sole

plates where required, and should be no more than 2.5 m apart.
(2) Ledgers must be fixed to standards with double couplers, the normal height of each lift being 2 m, which provides headroom for working on intermediate platforms where required.
(4) Transoms should be fixed to standards where possible with double couplers, intermediate transoms being fixed to ledgers with putlog couplers.
(5) Diagonal bracing should be fixed on all sides with plan bracing at the base and other levels where necessary for rigidity. Figure 6.27 shows the first lift of a tubular tower scaffold.
(6) Where the height of the tower is more than three-and-a-half times the shortest side it must be adequately tied.

Mobile Towers (figure 6.28)

Relevant details are similar to those for independent tower scaffolds, except for the base plates, which are replaced by lockable castors. Other requirements are as follows.

(1) The maximum height of internal towers is three-and-a-half times the shortest base dimension, while that for external towers is three times these dimensions.
(2) When the scaffold has to be moved, force should be applied near the base; do not pull or push it along while standing on a platform.
(3) Use only on ground that is firm and level.
(4) Brakes must be locked on when in use and the tower should be tied to the structure whenever possible.
(5) Only one working lift should be in use at a time.

Framed or System Scaffolds (figures 6.28 and 6.29)

These usually consist of metal H-frames constructed from patent welded units, which can be quickly interlinked to form independent or tower scaffolds. Each frame consists of two vertical members and one or two cross members. They are erected by joining two frames together with cross braces or ties and the height is increased by locating the next pair of frames over the spigots on the lower frames. See figure 6.29, which shows S.G.B. Sureframe scaffold, which, although a product of the 1960s, is still very popular, especially for support work. More recent developments in system scaffolding include, for example, Kwikstage and Cuplock, which are representative of the late 1970s.

With the Sureframe scaffold (figure 6.29) no couplers or fittings are necessary for the ledger and transom assembly, which carries the ends of abutting

Figure 6.27 First lift of an independent tower scaffold

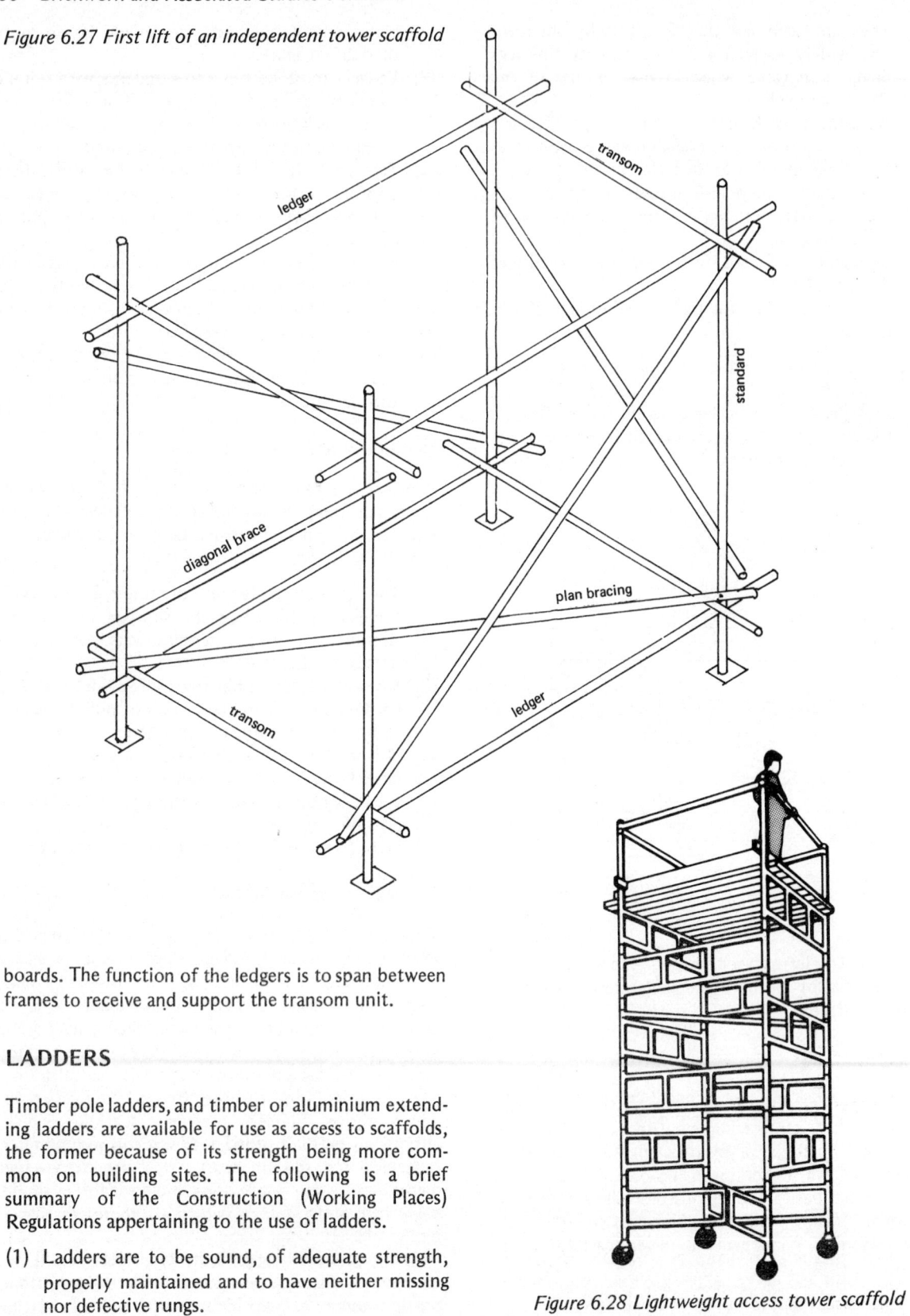

boards. The function of the ledgers is to span between frames to receive and support the transom unit.

LADDERS

Timber pole ladders, and timber or aluminium extending ladders are available for use as access to scaffolds, the former because of its strength being more common on building sites. The following is a brief summary of the Construction (Working Places) Regulations appertaining to the use of ladders.

(1) Ladders are to be sound, of adequate strength, properly maintained and to have neither missing nor defective rungs.

Figure 6.28 Lightweight access tower scaffold

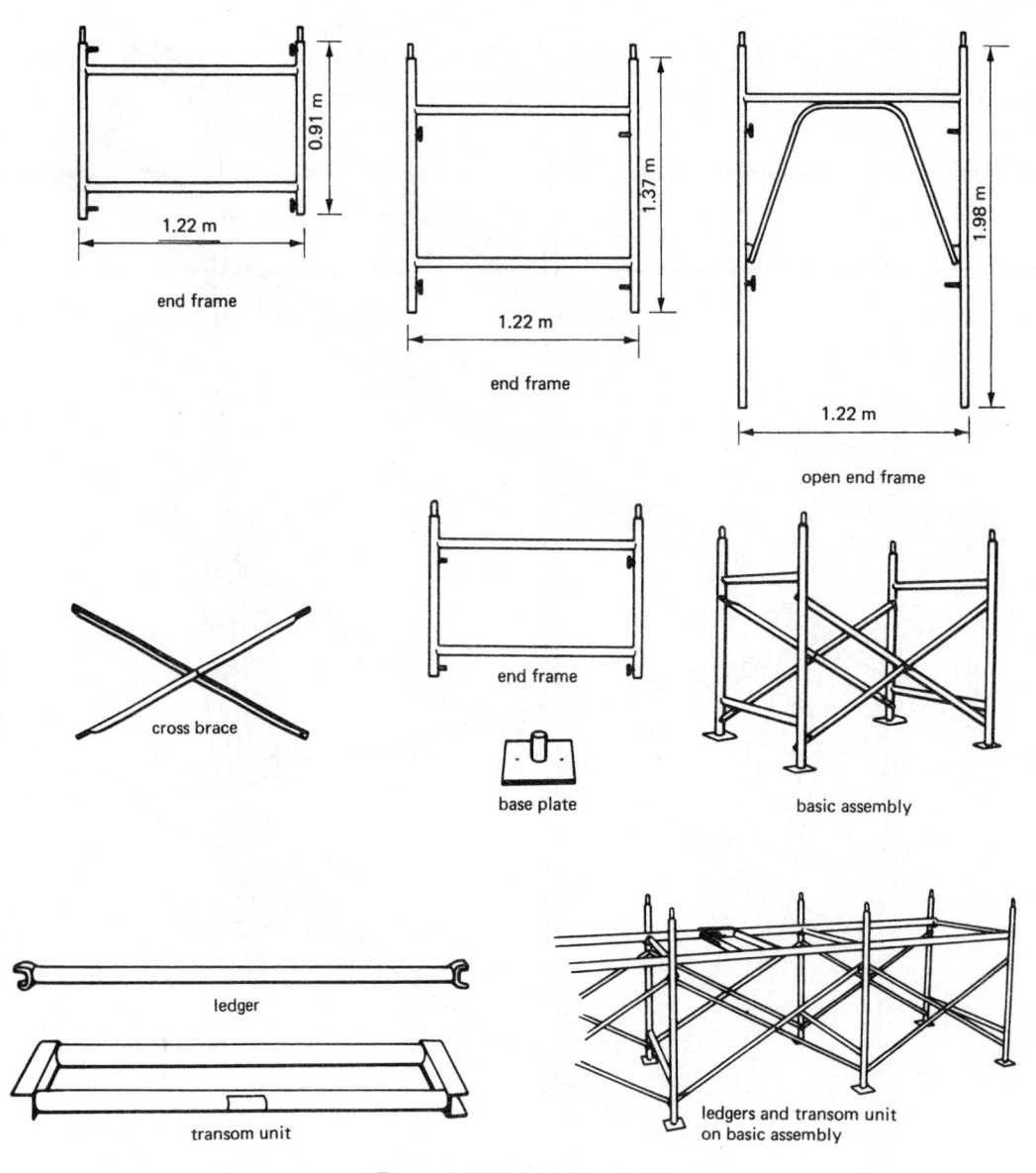

end frame

0.91 m

1.22 m

end frame

1.37 m

1.22 m

open end frame

1.98 m

1.22 m

cross brace

end frame

base plate

basic assembly

ledger

transom unit

ledgers and transom unit
on basic assembly

Figure 6.29 S.G.B.Sureframe

(2) Rungs are to be properly fixed to the stiles and must not rely on nails.

(3) Where possible ladders must be fixed at their upper end to the scaffold and must extend at least 1 m above the platform unless some other handhold is provided.

(4) Where top fixing is impracticable ladders should be fixed near the bottom and must not be allowed to sway or sag unduly.

(5) Both stiles must be equally and firmly supported and must not be stood on loose bricks or other loose packings.

(6) Landings must be provided every 9 m in height.

(7) Ladders must not be painted to hide defects.

Ladders should be inclined at an angle of or near 75°, commonly referred to as four up, one out (figure 6.30). While a suitable procedure is to rear the ladder against the scaffold it is better to prevent the overhang at the top from encroaching on the scaffold by rearing the ladder against an extended putlog or transom, which must be secured to the ledger with a double coupler (figure 6.30*b*).

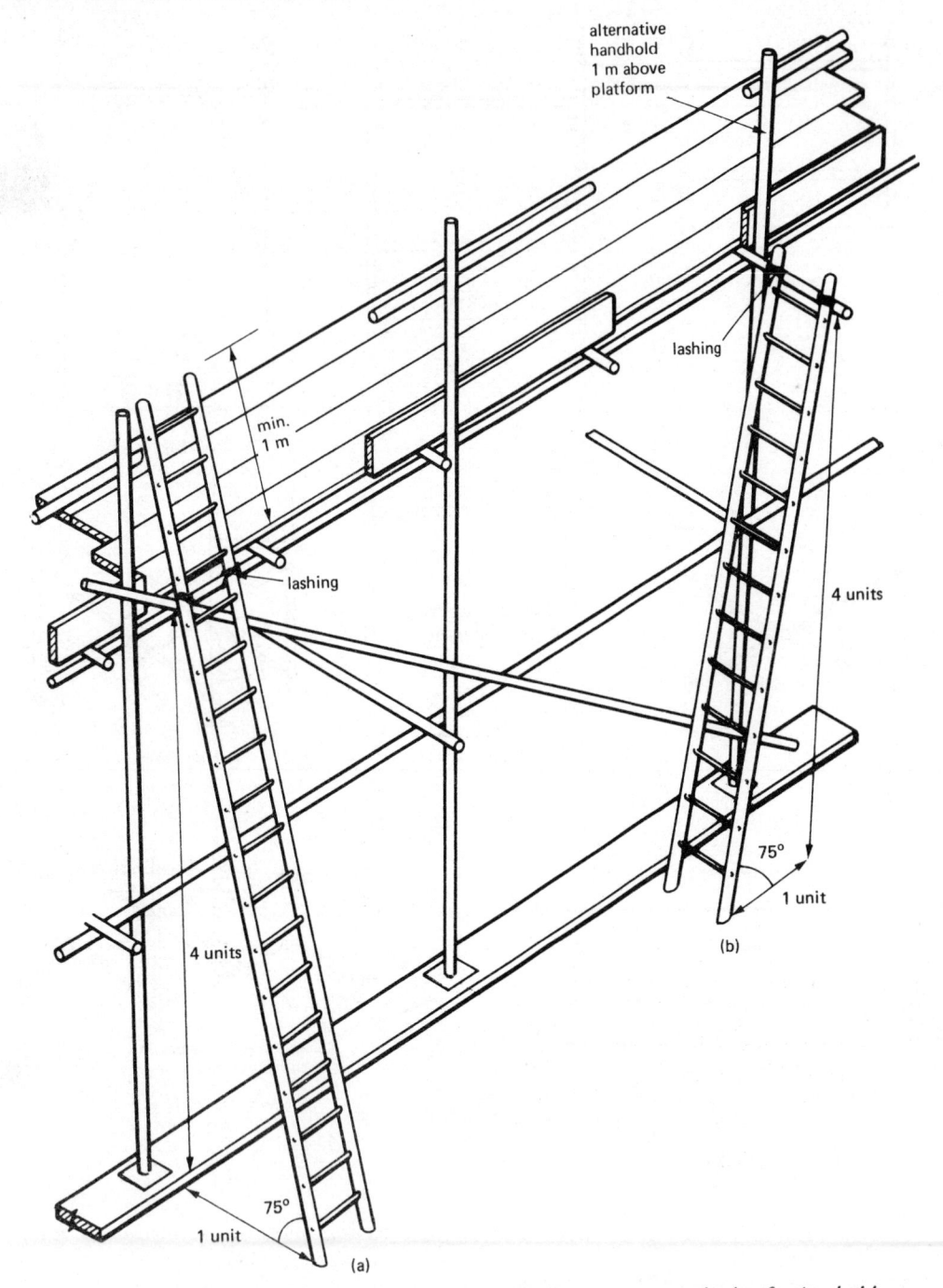

Figure 6.30 Isometric drawing of a putlog scaffold showing two methods of using ladders

LIFTING EQUIPMENT

The Wedge

This is a simple inclined plane, usually of timber, which is moved forward by a series of hammer blows while the body (the object to be raised) remains — as it were — in a fixed position. Figure 6.31 shows the forces involved, and a comparison of figures 6.31 and 6.32 makes it obvious that, although the amount of lift is less, the narrower the wedge, the less effort is required. Wedges are often used in pairs (figure 6.33)

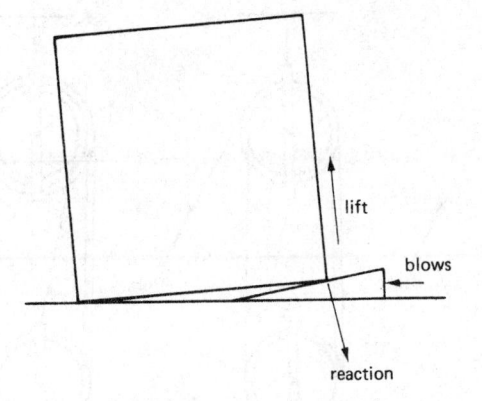

Figure 6.31 Use of narrow wedge

and a series of blows to each wedge will raise the object without tilting occurring.

The Screw Jack

This too is an inclined plane in spiral form, in which the lever is rotated about its vertical axis in order to raise the load, which is placed on the swivel head. The distance between the top of one thread and the top of the adjacent thread is known as the pitch of the screw. The length of the lever may be as long as 900 mm in some cases and it should be clear from figure 6.34 that one complete revolution of the lever will raise the swivel head a distance equal to the pitch.

With the screw jack little strength is required to lift heavy loads. An Acrow prop, for example, works on this principle, and although the lever is only about 225 mm long the lifting and supporting capacity is well known in the construction industry.

As is the usual case with lifting equipment, however, we are not getting something for nothing since the effort is applied at the end of the lever, and this has to be rotated through a large distance to raise the

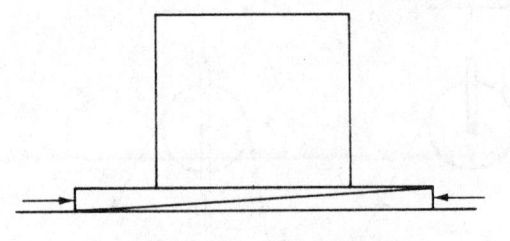

Figure 6.33 Use of folding wedges

load fractionally. For example, if the pitch of the screw is 5 mm and the length of the lever is 900 mm, the end of the lever moves through a distance of more than 5½ m to raise the load 5 mm; and with an Acrow prop, although the lever is only 216 mm long, it turns through a distance of nearly 1.6 m to raise the load 6 mm (dimensions converted).

Simple Pulleys

With a pulley it is possible for a man to raise an object several times his own mass, on to a scaffold, with the minimum of effort. The act of bending down and lifting a heavy object is difficult and may be dangerous, and it is much better to raise the object by heaving downwards on a rope. The simplest form of pulley is the gin wheel, which consists of a single wheel over which the rope is passed (figure 6.35a). The hook must be firmly secured to an extended putlog or transom with a figure-of-eight wire lashing turned at least five times round the hook and arranged so that the hook hangs 75 or 100 mm below the tube.

With an independent scaffold the support tube must be connected to both standards (figure 6.4), and with a putlog scaffold, the support tube should be

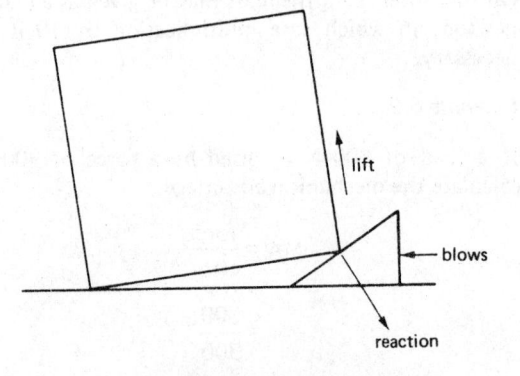

Figure 6.32 Use of steep wedge

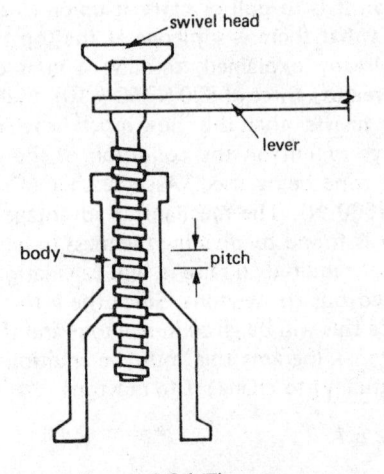

Figure 6.34 The screw

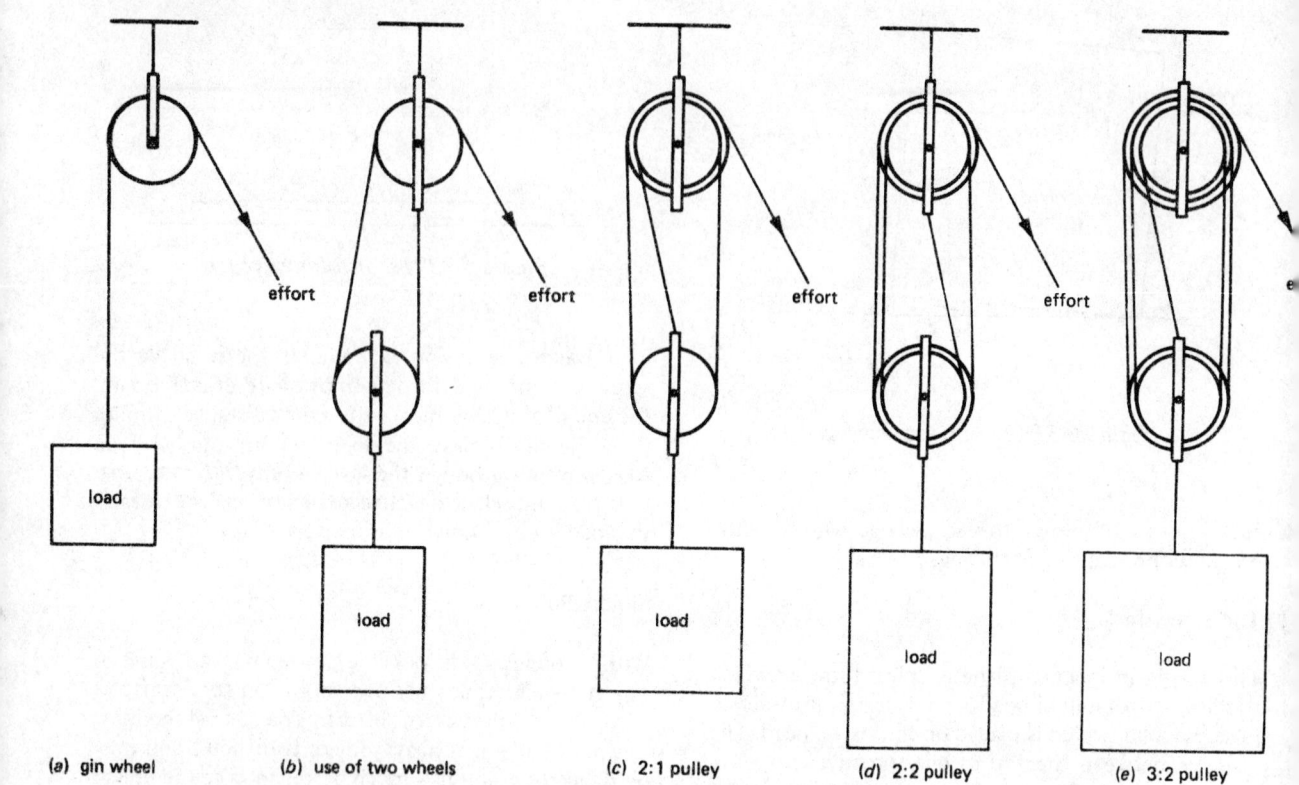

Figure 6.35 *Pulleys (the wheels are shown with different diameters so as not to obscure the ropes)*

connected to a standard and braced back to the level of mature brickwork. Support tubes to gin wheels should be placed as near to a positive tie as possible. This load is then attached to one end of the rope and the effort is applied at the other.

With a gin wheel a man cannot raise an object as heavy as himself and therefore the main reason for using this lifting appliance is that it is easier and safer to raise a heavy object by pulling downwards on a rope than it is to pull or carry it up on to a scaffold, provided that there is someone at the top to unload.

As already explained, to raise a mass of 50 kg, which creates a force of 500 N (50 x 10), the operative must be heavier than this; how much heavier depends to a large extent on the condition of the gin wheel and the rope being used. Assume that 60 kg is sufficient (600 N). The mechanical advantage (MA) of a pulley is found by dividing the mass to be raised by the effort required to raise it, and calculations should be carried out in newtons. Since the effort required is a force this will be given in newtons and if the mass is given in kilograms this must be multiplied by 10 (9.81 actually) to change it to newtons.

Example 6.1

If a mass of 50 kg can be raised by an effort of 600 N, what is the mechanical advantage?

$$MA = \frac{mass \times 10}{effort} \quad \begin{array}{l}\text{(to change kg to N)}\\ \text{(already in N)}\end{array}$$

$$= \frac{50 \times 10}{600}$$

$$= \frac{500}{600}$$

$$= 0.833$$

On the other hand the mass may be given as a load in newtons in which case multiplication by 10 is unnecessary.

Example 6.2

If a load of 800 N is raised by a force of 900 N, calculate the mechanical advantage

$$MA = \frac{force}{effort}$$

$$= \frac{800}{900}$$

$$= 0.88$$

The velocity ratio (VR) of a pulley is defined as the downward distance moved by the rope due to the effort compared with the upward movement of the load. It will be obvious that with a gin wheel the amounts of upward and downward movement are equal and, therefore, the VR of a gin wheel is unity.

Note The VR of a simple pulley can also be found by counting the number of wheels used in the system. The efficiency of a pulley is calculated by dividing the mechanical advantage by the velocity ratio and multiplying the result by 100. In the case of example 6.1

$$\text{efficiency} = \frac{MA}{VR} \times 100\%$$

$$= \frac{0.833}{1} \times 100$$

$$= 83.3\%$$

Using more Wheels

When a hanging pulley attached to a load is supported by two ropes (figure 6.35*b*) the force in each rope is equal to half the force exerted by the mass. Therefore, to balance a mass of 100 kg an effort of 500 N (50 kg x 10) would be required. The VR would be 2 since pulling 300 mm downwards would only raise the load 150 mm (also, two wheels are being used). In figure 6.35*c* there are three ropes supporting the load and, therefore, if the load to be raised had a mass of 150 kg the force in each rope would be 500 N (50 kg x 10) and thus an effort of 500 N would balance a load of 150 kg, which creates a force of 1500 N (150 kg x 10). To raise the mass a little extra effort will be required.

Example 6.3

Assume that the mass to be raised is 180 kg and the force required to raise this is 800 N (80 kg x 10). Calculate MA, VR and efficiency.

$$MA = \frac{\text{mass} \times 10}{\text{effort}}$$

$$= \frac{180 \times 10}{800}$$

$$= \frac{1800}{800}$$

$$= 2.25$$

(This means that a man can raise 2¼ times his own mass when using this pulley.)

VR = 3 (number of wheels in system)

$$\text{efficiency} = \frac{MA}{VR} \times 100\%$$

$$= \frac{2.25}{3} \times 100\%$$

$$= \frac{225}{3}$$

$$= 75\%$$

Consider figures 6.35*d* and *e*. While these are a little more complicated, still larger mechanical advantages can be gained with their use. In figure 6.35*e*, for example, a load having a mass of 250 kg can be balanced by a force of 500 N (50 kg x 10). This particular pulley has a velocity ratio of 5 (number of wheels used in the system).

The Scaffold Crane

This is loosely described as a powered gin wheel and it will raise a mass of up to 250 kg on to a scaffold platform. The crane is situated on and operated from the platform and the fixed slewing jib which is extended outwards must be adequately braced back to a standard before lifting starts. The load can only be moved laterally in a circle of fixed radius.

Elevators (figure 6.36)

Manual and hydraulic elevators are available that are either towable as a two-wheel trailer behind a car or lorry or can be mounted on a lorry chassis. They are very useful for building contractors, scaffolders, glaziers, slaters and tilers, etc. since suitably sized platforms are available for attaching to the elevator to carry anything from bricks to 6 m scaffold tubes. Figure 6.37 shows the top of an elevator transporting bricks.

The heights and angles to which elevators can be used vary greatly: those produced by Walter Somers, for example, vary in height from 8 to 30 m and in certain circumstances extensions to 40 m are possible. They can be erected by one man, and are ready for work to start in 10 minutes; with a man to load at the bottom and another to unload at the top, materials are very quickly transferred. Where materials are intended to be tipped at roof or other level this facility can be incorporated into the elevator (figure 6.38).

Figure 6.36

Figure 6.38 Combined tipping and rubbish bucket for tipping below and above

Hoists

A hoist consists of a horizontal platform which moves up and down vertical guides by a powered winch. The guides are normally tied back to the structure or scaffolding to provide stability.

The Construction (Lifting Operations) Regulations 1961, part V cover the requirements for hoists and these are summarised as follows.

(1) The hoistway must be protected by a substantial enclosure at all points where access is provided or where persons may be struck by moving parts.
(2) Safety gates must be provided at each landing for loading and unloading.
(3) An automatic brake is to be provided to support the platform in the event of failure of ropes or any other part.
(4) A safety device must be installed to ensure that the hoist cannot overrun its highest point
(5) The safe working load is to be clearly marked and this must not be exceeded except for testing. A hoist to carry persons is to state the maximum number to be carried.
(6) A notice is to be placed on the platform stating that persons are not to be carried unless the hoist is so designed (part VI 48).
(7) The hoist is to be operated from one position (not from within unless part VI 48 is complied with) and if the operator cannot clearly see the platform arrangements for suitable operating signals must be made.

Figure 6.37

Mobile Hoists

With this type of hoist the mast can be quickly lowered by two or three men and moved on two pneumatic tyres. It was developed mainly for housing, and models range from those that can lift 250 kg to a height of 4 m, to the largest, which can raise 1 tonne to a height of 5 m. The mast is extendable in some cases, but over 6 to 8 m — according to the model — the mast must be tied to the building, in which case it can no longer be classified as mobile. Mobile hoists are very useful for serving two or three houses in close proximity since in this type of work the rate at which materials are used is usually relatively slow.

Fixed Hoists

As the name implies, this type of hoist remains in a fixed position throughout a contract and sizes vary from ½ to 3 tonnes capacity, which can be raised to heights of about 150 m. The platforms of these hoists can be side-slung as in the previous case, or centre-slung, in which case a tubular scaffold tower is erected to enclose it. To stabilise a fixed hoist it must be tied at regular intervals to the building.

7
MAINTENANCE, REPAIR AND FIXING EQUIPMENT

The construction industry provides its craftsmen with a greater variety of work than any other industry. In the construction of new buildings, which is classified as new work, employment is provided for approximately 25—35 per cent of the labour force. In maintenance, repair and alteration, work is constantly provided for the majority of building trade workers.

In the work of maintenance and repair we find that situations, buildings and materials are never identical, and techniques that are suitable for a particular situation may not be universally applicable because so many other factors have to be considered. The bricklayer craftsman should, therefore, be equipped with the skills that will enable him to cope with the many different problems and varying situations in which he may become involved.

REPLACING DEFECTIVE FIREBACKS

All types of fireback may become defective because of the following causes

(1) abrasion and resistance to heat, which is termed normal wear and tear
(2) movement of the surrounding infill material, the structural hearth, tiled hearth and tiled surround: if any movement caused is not accommodated it will result in damage to the fireback.

It is recognised that sectional firebacks are capable of resisting movement better than the single unit or one-piece fireback, and obviously the four-section fireback is more capable than one comprising two sections.

When the craftsman is required to renew an existing defective fireback, it is important to inspect and determine the area of damage and the cause of the trouble.

Figure 7.1a shows areas of disintegration in the fireback, with spalling, laminating and sometimes fractures originating from the worn areas. Where only slight cracking and a minimum amount of spalling is found, repair can be effected by pointing with patent fire-cement, otherwise complete reinstatement is required.

Figure 7.1b shows a very large crack occurring at the centre of the fireback, running from top to bottom. This is caused by lateral pressure on the fireback, resulting from the absence of a movement joint around the back of the fireback, with consequent pressure from the infill material.

Figure 7.1c shows a large horizontal crack occurring about half way up the fireback. This is caused by uneven pressure on the fireback, and is due to movement of the infill material and to the absence of a movement joint between the fireback and the tiled surround.

Figure 7.1d shows a deep vertical centre fracture and also fractures in the side cheeks. The causes of this defect are the total absence of any movement joints, badly placed infill material and sometimes the wrong type of infill material.

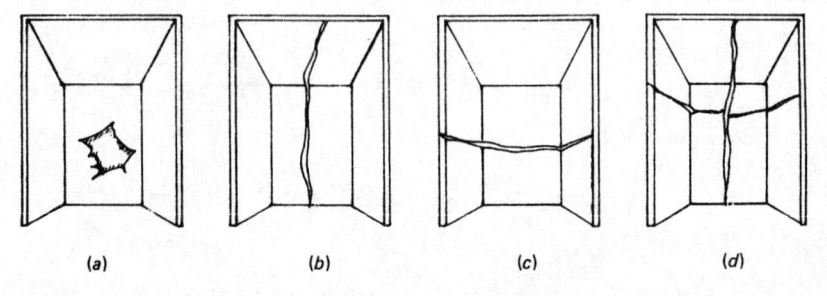

| (a) | (b) | (c) | (d) |

Figure 7.1 Defective firebacks

Removing the Fireback

Where the fireback consists of a one-piece unit, removal can easily be effected with the lump hammer and an 18 mm-diameter cold-steel chisel at least 250–300 mm long. The cutting-out operation should be started around the central fracture and the side cheeks can be taken out after the centre of the back has been removed.

It should be noted that before any of the above operations are begun, it is essential to provide complete protection for the tiled surround and hearth. This can be done with sacking over the hearth and drapes over the tiled surround, which will prevent any chipping or spalling of the tiles should they be struck by particles from the fireback.

Removing the Infill Material

Removing the infill material may require the craftsman to use a lightweight percussion drill with a chisel end, or the lump hammer and a 25 mm-diameter cold-steel chisel at least 450 mm long.

After Removing the Infill

After all the infill material has been removed and the area within the fireplace opening has been completely cleaned out, a hole 225 x 225 mm should be cut out above the tiled surround and structural lintel (figure 7.2). The hole is formed to ensure that making good can be completed above the top of the new fireback, and at the position of the throat.

Fixing the New Fireback

Before determining the type of fireback to be inserted, it is essential to recognise

(1) the type of solid fuel that is to be used
(2) the amount of abrasion to be expected
(3) the type of heat resistance necessary.

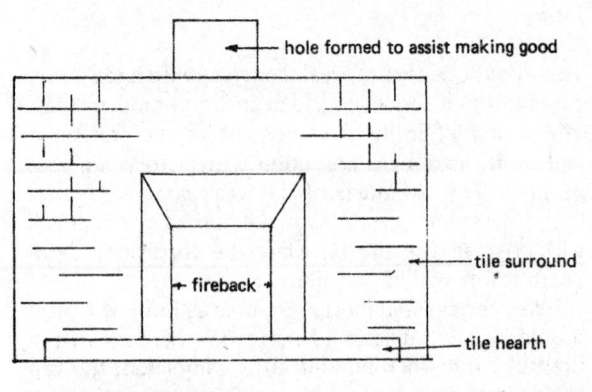

Figure 7.2 Method of making good the infill behind and above fireback

The above factors then determine the type of fireback to be used and the material that forms the unit. If the single or one-piece fireback is used it should be inserted through the opening and then manoeuvred until it stands upright within the fireplace opening.

Before the fireback is placed in its position at the back of the tiled surround, two movement joints should be formed on each side of the front of the fireback, or on the back of the tiled surround. These movement joints consist of a length of asbestos rope, 25 mm in diameter, which should have been soaked in water glass and cut to a length equal to the height of the fireback. These will then form the movement joints between the front of the fireback and the back of the surround (figure 7.3).

The fireback is then positioned and a further movement joint is formed around the entire area of the fireback. This joint is formed by wrapping corrugated cardboard around the positioned fireback. When used as a movement joint this material will later disintegrate with heat, thereby leaving an open joint between the infill and the fireback, which will allow movement to take place (figure 7.3*b*).

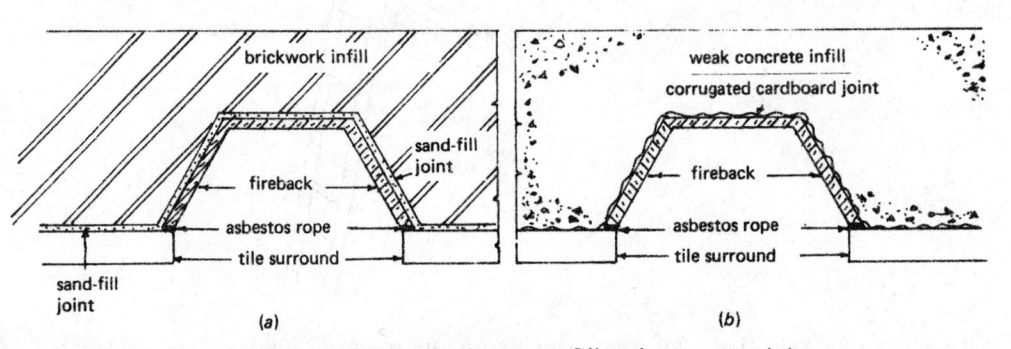

Figure 7.3 Plan of firebacks showing infill and movement joints

Infill

This should consist of weak concrete with a ratio mix of 1:4:10, or bricks, bedded in lime—sand mortar. The choice of infill is usually left to the craftsman and both materials are quite satisfactory for the purpose. The advantage of the weak concrete is that it can easily be placed over the top of the fireback with a small fire shovel, while the corrugated cardboard is retained in position.

When bricks and mortar are used as infill, the infill should be kept at least 12 mm from the back of the fireback, and when building-in is completed, the gap is then filled with sand (figure 7.3a) to provide the movement joint.

Use of the Preformed Hole

The hole previously formed above the structural lintel can now be used to build up and form the throat. This is done with bricks and mortar formed on top of the infill, within the fireplace opening. When this operation has been completed the hole can then be made good and replastered to the existing plasterwork.

DEFECTIVE CHIMNEY STACKS

Work Before Inspection

A suitable type of ladder should be erected and placed in position to provide easy access to the roof. Because materials may be taken on and off the roof, it is advisable to use a pole-sided type, which should be secured at the required height above the eaves and at the correct angle (figure 7.4). Crawling boards or crawling ladders are then placed in position on the roof to allow inspection of the chimney stack.

Inspection

This can now start with the chimney capping and then the entire brickwork of the stack. The ability to recognise and determine the following factors is a prerequisite for inspection

(1) the condition of the capping and the capping material
(2) the type and condition of the bricks and mortar used for the stack
(3) any visible fractures in the brickwork or flaunching

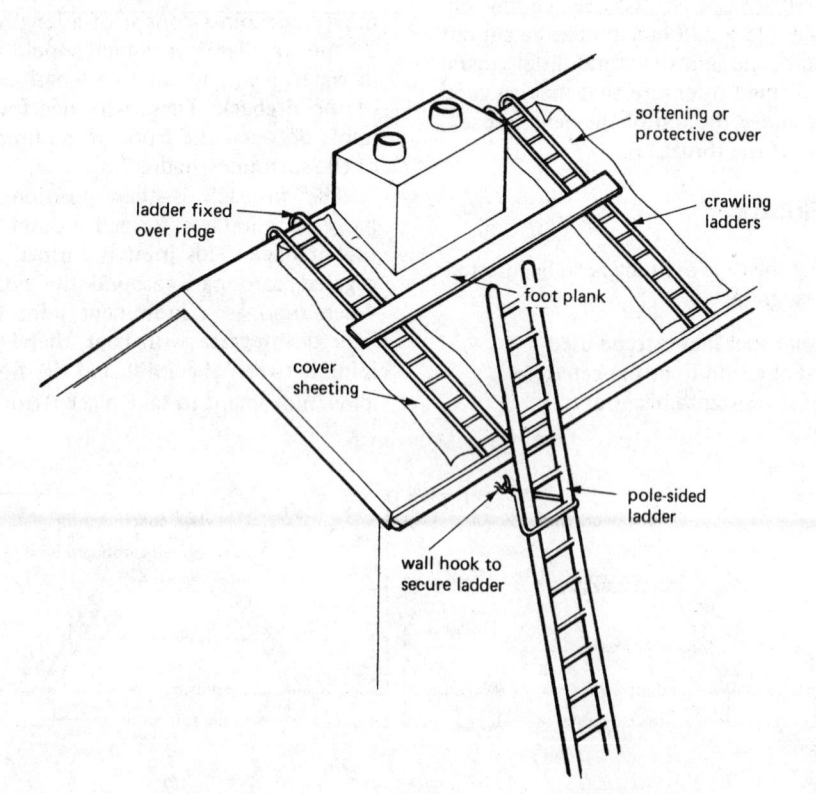

Figure 7.4 Requirements for inspection of chimney

(4) whether the damp-proof course is effective, whether there are flashings and soakers, and in what condition.

Faults and Failures

The following faults and failures are recognised as being common and are often found on existing houses and old buildings

(1) leaning chimney pots
(2) loose capping brickwork
(3) cracked and loose flaunching
(4) fractured stacks, showing visible cracks
(5) spalling and lamination of the stack bricks
(6) deterioration of the mortar joints
(7) leaning chimney stacks
(8) ineffective d.p.c.

The first seven failures are often caused by sulphate attack and may be accelerated by the ineffectiveness of the chimney capping where it has insufficient projection or is badly weathered and formed with weak materials. Fault 8 can be considered to be a contributing factor to some of the other failures.

Conclusion of the Inspection

Where the condition of the stack is poor, and flue liners are obviously not present, the entire stack should be taken down to three to four courses below roof level and rebuilt with suitable new material, with flue liners inserted during construction.

Protection

Before any work is carried out on the chimney stack, the fireplaces below should be inspected and sacking should be inserted at the fireplace throat to prevent debris entering the room. All roof work should be covered and protected in the area of the stack; precautions should also be taken to protect any work, or people below roof level.

Equipment

Where the chimney is of considerable height, scaffolding must be erected around the stack. Timber is normally used for roof work, although other methods can be used to provide the necessary working platform (figures 7.5–7.7).

Taking Down for the Stack

This should be done with considerable care, the dismantling starting with the chimney pot and flaunching, then the stack itself. All materials should be removed and taken down to ground level, then placed in a position where they will not impede any building operations and also where they can easily be removed from site.

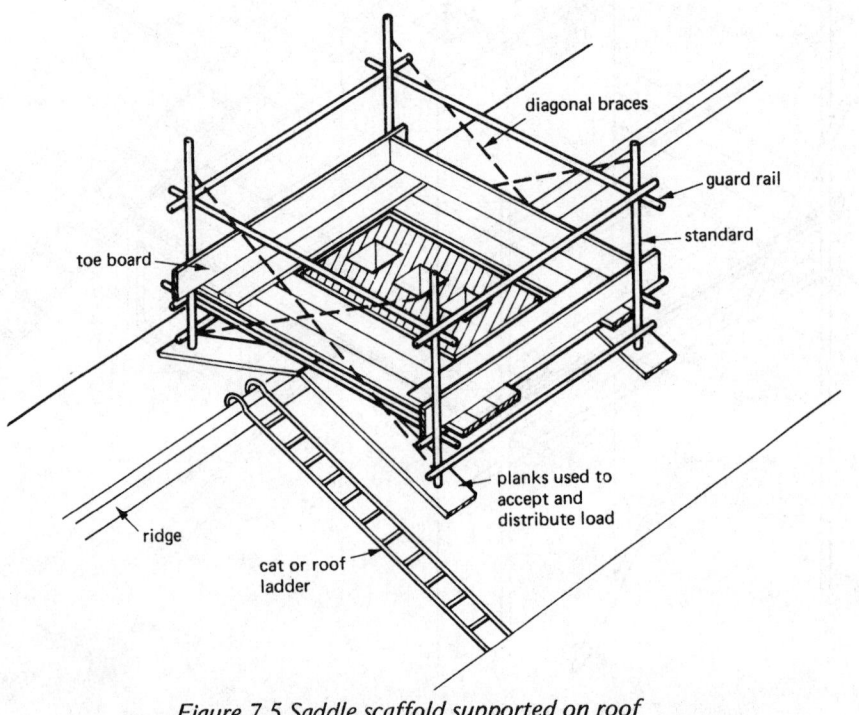

Figure 7.5 Saddle scaffold supported on roof

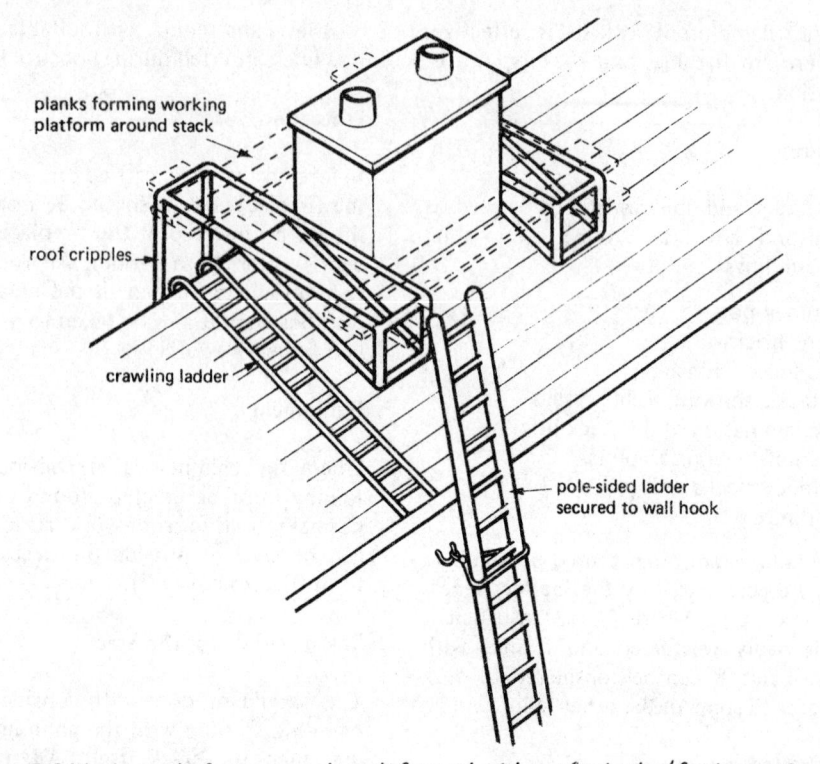

planks forming working
platform around stack

roof cripples

crawling ladder

pole-sided ladder
secured to wall hook

Figure 7.6 Working platform around stack formed with roof cripples (for inspecting only)

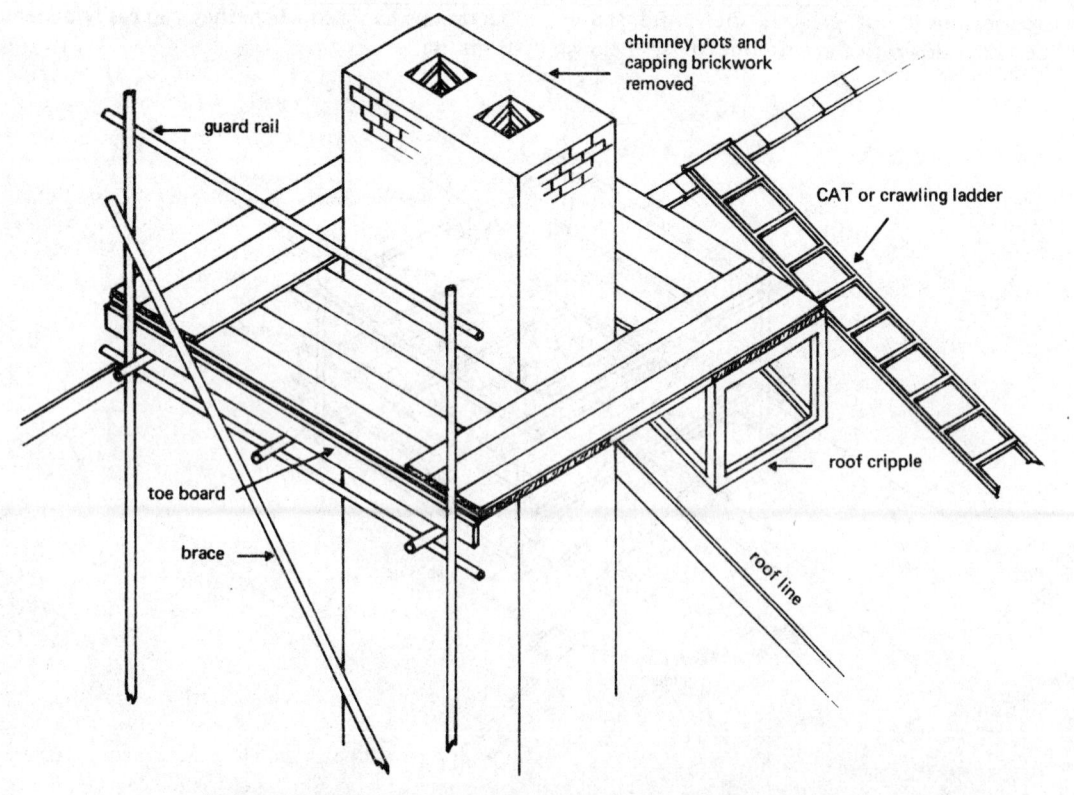

chimney pots and
capping brickwork
removed

guard rail

CAT or crawling ladder

roof cripple

toe board

brace

roof line

Figure 7.7 Working platform around external stacks on gables

Rebuilding the Stack

Materials

The type of bricks and mortar should be determined with considerable care, taking into account the previous failures, the necessity to combat sulphate attack, prevent weather penetration and blend in with the appearance of the building. Obviously, flue liners must be inserted at the position where rebuilding is begun.

Procedure

At the required position below roof level, rebuilding is begun. When the work reaches 150 mm above the lowest point of intersection between roof and stack, a tray d.p.c. should be inserted; the mortar joints around the base of the stack above roof level should be raked out to accommodate the flashings. The flue liners should continue to be inserted and built-in up to chimney-pot level, and it should be ensured that jointing of the liners is carried out with suitable material.

At capping level, adequate protection should be provided, oversailing courses should be formed or a precast concrete slab used to form the entire capping as one unit. The chimney pot can then be inserted and fixed. The pot should be selected according to the following requirements

(1) the type of appliance used in the fireplace opening
(2) the type of fuel used
(3) the need to complement the chimney stack and building below.

After rebuilding is completed, the sacking can be removed and the flue examined for draught by lighting a low fire. If the results are good, the scaffolding above roof level can be taken down, and the roof completely cleaned off, with all protecting covering removed, and the building waste taken away.

DEFECTIVE BRICK-ON-EDGE WINDOW SILL

It should be recognised that window sills that are constructed with bricks and mortar can become defective as a result of any of the following

(1) poorly selected bricks
(2) weak or poorly mixed mortar
(3) large mortar joints
(4) poor construction due to bad workmanship
(5) water remaining on top of the sill
(6) insufficient projection.

To construct a good weather-resistant brick sill, it is important that the following considerations are complied with.

(1) The bricks used for the sill should enhance and complement the building, but should be as dense and non-absorbent as possible.
(2) The mortar used to bed and form the sill should be of the same density as the bricks.
(3) Mortar joints should be correctly filled and should not exceed 6 mm in thickness.
(4) The amount of fall provided for the top surface of the sill should be 6 mm per 100 mm of sill surface.
(5) The projection of the sill beyond the brickwork face should be at least 50 mm.

Inspection of Sill

The sill should be inspected and examined to determine the cause of the deterioration.

Removal of Sill

Removing a badly defective sill should not cause any problems for the craftsman and it can be effected with the normal bricklayer's tools.

Cutting out should start at each end, taking out the first two bricks from each end; the third brick from the end on each side should be left to support the sill, and the bricks between should then be taken out. Temporary packings are then inserted, and the two supporting bricks are removed (figure 7.8).

After this operation has been completed the area of brickwork below the sill, termed the apron, should then be examined and, where repointing is necessary, raking out and cleaning should be carried out at this stage.

Reinstatement of Sill

The selected bricks should be placed on a flat surface and checked for alignment, with the correct joint allowance formed between the bricks; the length of brick can then be marked on the top surface, a gauge staff can be formed and the brick can then be cut to the required length (figure 7.9).

Having cut the bricks, dampen the bedding surface and where repointing is required dampen the area of the apron. Bed, joint and fix the first two bricks at each end, checking for fall, projection and gauge. A line can then be fixed to top or bottom arris, depending whether the sill is above or below eye level. The temporary packings should then be removed and the operation of bedding and fixing the brick on edge continued. This should be carried out by working

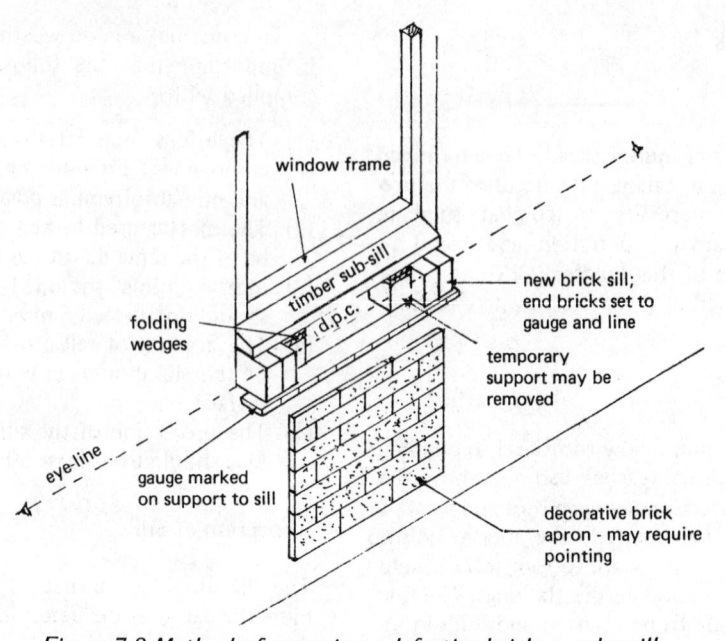

Figure 7.8 Method of renewing a defective brick-on-edge sill

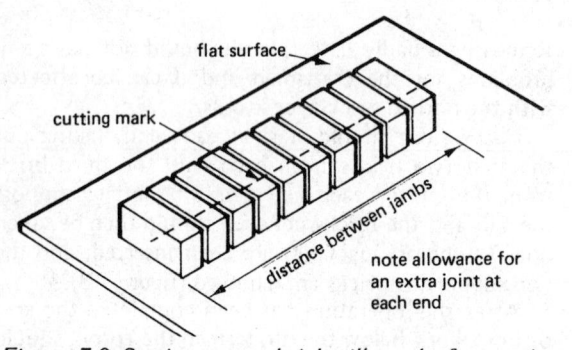

Figure 7.9 Setting out a brick sill ready for cutting

from each end to the centre, checking for gauge as the work proceeds. The brick sill can be jointed after completion and the pointing of the apron also completed.

Where a wooden sub-sill is attached to the window the joint between the wooden sill and the brick sill should be pointed with a mastic compound.

REPOINTING OLD BUILDINGS

Before any operations are carried out on the face of the building, it is important that a close inspection of the walling should be carried out.

Inspection of the Building Face

It is very important to determine the position of the walling and to recognise the amount of weathering or atmospheric pollution that the walling is required to withstand; the amount and type of deterioration of the wall surface, the number of laminated bricks and the condition of the mortar joints should all be carefully determined and assessed. Factors that should also be considered are as follows.

(1) Where staining of the brickwork face is evident, what is the type of staining, and how was the staining caused?
(2) Can the brickwork face be cleaned and what method of cleaning would be most effective?
(3) Is protection necessary, and how can it be ensured?
(4) Where bricks are defective, can they be removed and similar bricks obtained and then replaced?

It is very important that the above factors should be considered and a positive conclusion reached in each case before any form of work is started.

Scaffolding

Where scaffolding is required for defective brickwork to be repointed, cleaned and reinstated, careful consideration should be given to the type of scaffolding that will be most effective. For this type of work, scaffolding should be completely safe and should comply with the Construction Regulations. The main functions of scaffolding are as follows.

(1) Fatigue for operatives using the scaffolding should be reduced to a minimum.
(2) It should be possible for all work above ground level to be carried out in an economical manner.

(3) Cleaning operations should not be impeded and the scaffolding should provide protection for the work, and also for persons who are at ground level.

Protection

Protective measures should be carried out before any form of work starts on the building face. They may consist of masking out all existing mastic pointing, glass, paintwork, timber and decorative features. Masonry and ironwork should also be protected.

Cutting Out and Reinstating Defective Bricks

This operation should be begun at the top of the building and worked down to ground level. The bricks on each lift should be cut out and replaced before work is begun on the lift or platform below.

Repointing

Before this operation is begun it is often advisable, when large structures are involved, to point panels at the base of the walling. These are termed sample panels and they are used by the architect to determine the type of mortar that will be the most suitable for the building. Whenever possible sample panels should be at least 1.0 m^2 and labelled with

(1) the type of sand and cement used
(2) the cement–sand ratio
(3) the water content
(4) the type and amount of colouring used.

Panels should be pointed and viewed after a minimum period of 72 hours, otherwise a complete appreciation of the panel is not possible.

The type of pointing joint is also selected by the architect, and it is determined by

(1) the weather and degree of exposure
(2) the condition of the brickwork face
(3) the decorative requirements
(4) the need for economy in the pointing operations.

Obviously the sample panels are formed with the pointing joint selected, which then allows its qualities to be appreciated.

Cleaning Brickwork before Repointing

Brickwork staining is often caused by external sources although it can also be due to salts in the bricks or mortar. To remove stains on brickwork requires considerable knowledge, skill and care, otherwise the use of wrong techniques or materials may cause permanent damage to the entire face of the walling.

Removing Stains from Clay Brickwork

Oil Stains   Sponge or poultice the area with white spirit, carbon tetrachloride or trichlorethylene. Where staining is severe, use several applications.

Efflorescence   Allow weather to take its natural course, but brush off with a fibre brush when the efflorescence is at its maximum. After a reasonable period wash down each week for a period of one month and allow the walling to dry out before further wetting is contemplated.

Paint Stains   Apply a patent paint remover as instructed, or use a solution of trisodium phosphate, 1 part to 5 parts of water (by weight). Allow the paint to soften and remove with a stiff fibre brush, washing down afterwards with soapy water.

Mortar Stains   When possible use a softwood scraper and wash down with a diluted solution of hydrochloric acid, 1 part to 10 parts of water (by volume).

Lichens and Mosses   First brush off with a stiff fibre brush, then use a patent moss killer as instructed, or a solution of zinc or magnesium silicofluoride, 1 part to 40 parts of water (by weight).

Rust or Iron Stains   First wash down with a solution of oxalic acid, 1 part to 10 parts of water (by weight). If the brown staining does not respond it is probably a manganese stain.

Manganese Stains   Brush down the staining with a solution of 1 part acetic acid, 1 part hydrogen peroxide and 6 parts water (by volume). Apply a second application only after a period of 3 days.

Lime Stains   Treat as for mortar stains.

Smoke or Soot Stains   Use a fibre brush and brush down gently. Apply a wash of household detergent, and where staining is heavy use trichlorethylene as a poultice.

Tar or Bitumen Stains   Use a stiff fibre brush and scrub down with an emulsifying detergent. When the area is dry it may be necessary to apply a paraffin-soaked sponge.

All brickwork should be completely washed down whenever any forms of acid have been applied. This may cause efflorescence but it will only be shortlived.

Commercial Cleaning

This is only economical where large areas of walling are to be cleaned, otherwise it may be very expensive in labour and material costs. Methods used for stone-work are steam, grit or sand-blasting, although the latter is only used as a last resort because of its effect on the surface of the walling material. Before steam, grit or sand-blasting is used, it is advisable to apply the methods on a sample panel before selection is made.

The above methods should only be considered for brickwork when it is judged impossible to obtain a clean surface with water and chemicals.

ALTERATION TO WALL LENGTHS AND THICKNESSES

Existing walls can be increased in length by any one of the following three methods

(1) cutting out block indents
(2) cutting out toothings
(3) forming a slip or vertical joint.

Choice of Method

The method of extending the length of a wall is determined by the following factors

(1) the situation of the wall, that is, whether it is external or internal, and whether the walling is facework or commons
(2) the requirements of the designer
(3) the materials used for the existing wall
(4) whether differential movement is anticipated.

Methods

Block Indents

These are formed by cutting out indentations in the end of the existing wall. The indents should be at least 100 mm in depth and in blocks of odd numbers, that is, of three or five courses. The maximum depth of any block is five courses; with this method the indent accommodates an even number of tie bricks, two or four (figures 7.10—7.12).

When cutting out the indents, it is advisable to start at the top and work downwards, thereby preventing the tails of any bricks above from snapping off (figure 7.10).

Block indents are normally used to extend the length of internal brick walls where accurate and normal bonding arrangements are of secondary importance, but where adequate strength from bonding-in can be obtained.

Building-in Block Indents   At the first course of the indent it is good practice to insert a wall tie or reinforcement (figure 7.12) to provide additional strength.

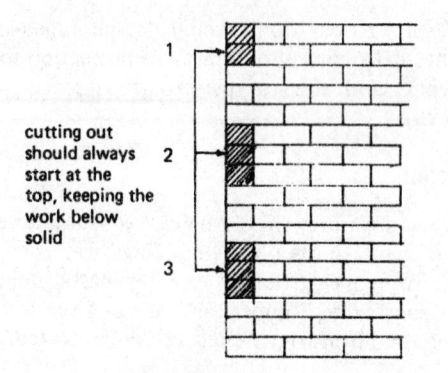

cutting out should always start at the top, keeping the work below solid

Figure 7.10 Preparing for block indents

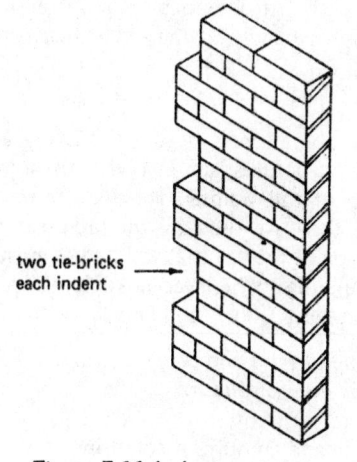

two tie-bricks each indent

Figure 7.11 Indents cut out

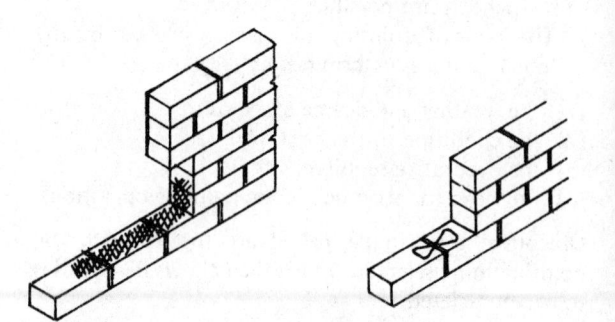

Figure 7.12 Using reinforcement to strengthen the tie when increasing wall lengths with block indents

The indent is then built up with considerable care being taken to caulk up the top bed joint in the indent with semi-stiff mortar. Obviously, if the existing walling is dry or dusty it should be brushed down and damped before joining up takes place.

Toothings

This method consists of cutting out every alternate brick at the end of the existing wall. The depth of the indent or toothing should be 56 or 110 mm, depending on the bonding arrangement of the existing wall and the new walling to be attached. Again it is essential to start cutting out the indents at the top and to work downwards, to prevent breaking off the projecting tails of the existing bricks (figure 7.13). When

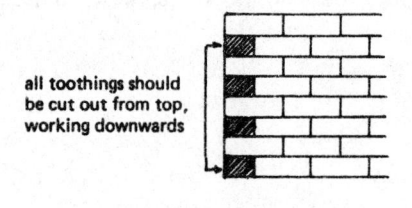

all toothings should
be cut out from top,
working downwards

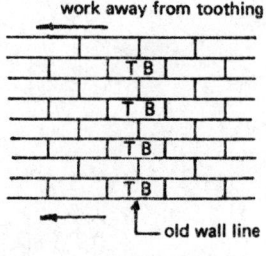

work away from toothing

old wall line

Figure 7.13 Increasing wall lengths by the toothing method

cutting out is completed, the toothings should be brushed out and if they are dry, damping will be required to ensure good adhesion between mortar and bricks.

The toothing method is extensively used on all external facework or wherever the bonding arrangement must be seen to be continuous throughout the length of walling (figure 7.13).

Building-in Toothings   Before any building work is started it is essential to ensure that the following recommendations are carried out.

(1) The dimensions of the new bricks should be the same as those used in the existing wall.
(2) The mortar should be the same density, texture and colour as the mortar in the existing wall.
(3) The toothed ends on the existing walls should be checked for vertical alignment, thus avoiding cutting to provide the correct bonding arrangement.

When building-in to toothings, line and pins should always be used to ensure horizontal alignment. Each

tie brick must always be inserted first and the course of bricks continued from this. The tie brick should never be inserted last, otherwise caulking and pinning up cannot be carried out satisfactorily. The caulking up should be completely solid and carried out with a semi-stiff mortar.

Slip Joint

This is sometimes referred to as a butt joint. The method consists of simply forming a vertical mortar joint, minimum 12 mm, at the end of the existing wall and starting the new work from this position. Often an open joint can be formed or butyl rubber or polysulphide compounds can be inserted. Both can have a mortar joint applied on the face later but the former method of leaving an open joint requires a movement joint to be gunned-in when the walls are completed (figures 7.14 and 7.15).

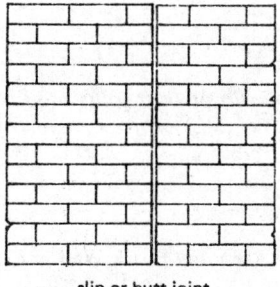

slip or butt joint

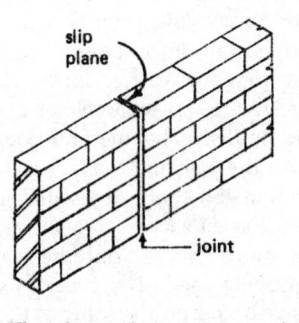

slip
plane

joint

Figure 7.14 The slip or butt joint used in brickwork

Slip joints are used where there is little knowledge of the existing foundation structure, or where unequal settlement is a possibility and differential movement may be expected.

It is important that coursing through from the existing brickwork is carried out. All the new work should again start from the vertical joint position to the opposite end of the new wall.

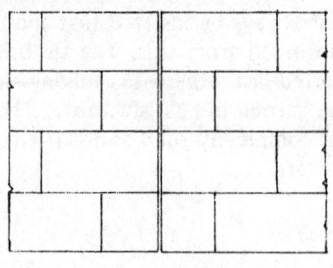

slip or butt joint

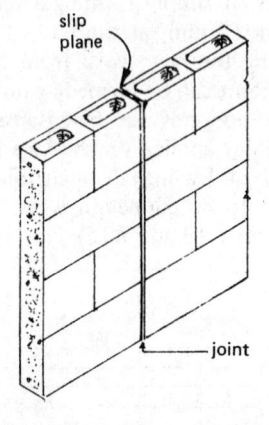

Figure 7.15 The slip or butt joint used in blockwork

Treatment for Extending Block Wall in Length

Block walls can be increased in length by any of the three methods previously mentioned, but where indents are to be cut into the end of the existing block wall, it may be easier, more accurate and economical to use a portable electric saw with a masonry blade or disc. This method does not affect the stability of the existing wall as often happens when heavy percussion tools are used. The procedure for building-in the blockwork is the same as for brickwork; the mortar should only be as dense as the walling blocks being used. The blocks used to form the new walling should be the same as those used for the existing wall; where this is not possible, joining up should be carried out with the slip joint.

INCREASING THE THICKNESS OF EXISTING WALLS

Existing walls can be increased in thickness by block bonding. This is considered the most practical method of tying two walls together to increase the thickness and also obtain the maximum amount of stability.

The method consists of cutting indents or recesses into the face of the existing wall; the indents should not exceed 100 mm in depth, while the length should be 225 mm and the height of the recess should be three courses (figures 7.16). The indents should be placed diagonally at 45° to the horizontal and cover the entire face of the wall (figure 7.17). Cutting out is performed with the normal bricklaying tools and also the portable electric saw with a masonry disc. Before building operations are begun, all indents should be brushed completely free of dust and damping down should be carried out when necessary.

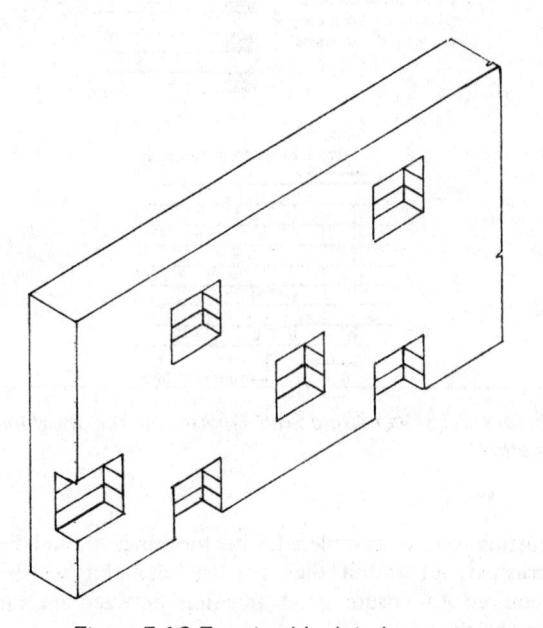

Figure 7.16 Forming block indents

Building Up the New Wall

Considerable care should be taken in the setting-out operations. The bond for the new facing wall should be arranged to ensure that header-tie bricks occur in each block indent. Whenever English bond is used for the facing wall there should be four tie headers per indent (figure 7.18). Flemish bond permits only two tie headers per indent (figure 7.19).

Caulking up at the top of each indent should take place to ensure the completion of the tie. It is good practice to use horizontal reinforcement in the new facing wall; wall ties should be used in all block indents.

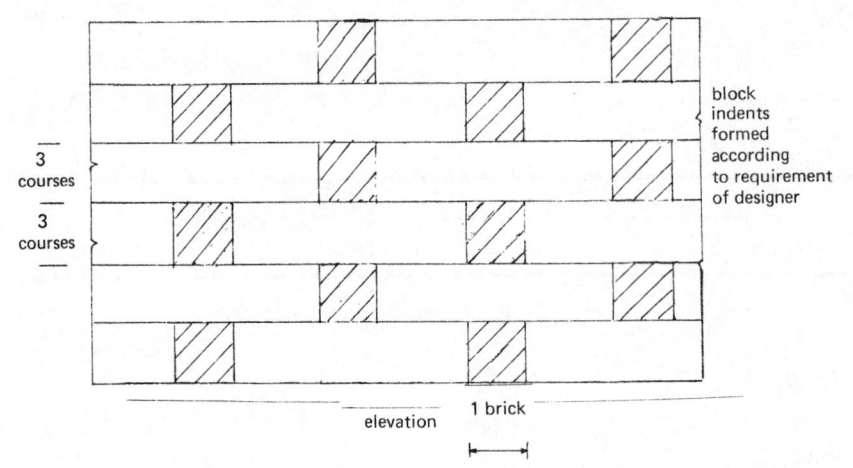

Figure 7.17 Increasing the thickness of walls by block bonding: elevation showing the position of indents

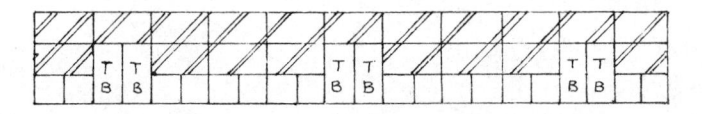

Figure 7.18 Increasing the thickness of walls by block bonding: plan of tie course for English bond

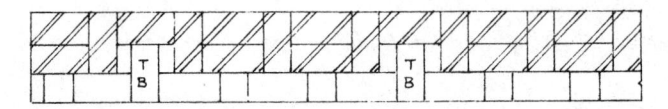

Figure 7.19 Increasing the thickness of walls by block bonding: plan of tie course for Flemish bond

FIXING CANTILEVER BRACKETS

Cantilever brackets can be fixed to solid or cavity walls of bricks or blocks. Before any operations are started it is necessary to inspect the walling because this often determines the types of tool to be used, the materials required and the type of temporary fixing equipment necessary.

It is important to recognise the condition of the walling and to determine whether it is of solid or cavity construction.

Procedure for Fixing a Line of Four Cantilever Brackets

First check whether the floor is level because this may alter the height of the brackets, depending on the designer's fixing requirements. A line should then be formed on the face of the wall at the required height. This is done with a chalk line, level and possibly a straightedge. The position of each bracket is then marked at centres on the chalk line.

Cutting-out operations can now be started. The cantilever brackets must be inserted into the wall to a depth of at least 100 mm to become effective, therefore it is essential to obtain a clean cut when cutting to form the soffit at the top of the bracket because upward movement cannot be allowed.

The dimensions of the holes should be not more than 28 mm on either side of the bracket. Whenever possible the bottom of the bracket should rest on the brickwork, which will ensure improved stability for the bracket and make fixing easier.

An assembled timber jig should now be set up, either by the carpenter and joiner, or the bricklayer craftsman himself. The jig should be adjustable to ensure that it is correct and level; the top rail should also be adjustable to ensure a level line for the brackets, therefore it is necessary to use folding wedges in both positions, as in figure 7.20. The jig should finally be positioned for alignment of the

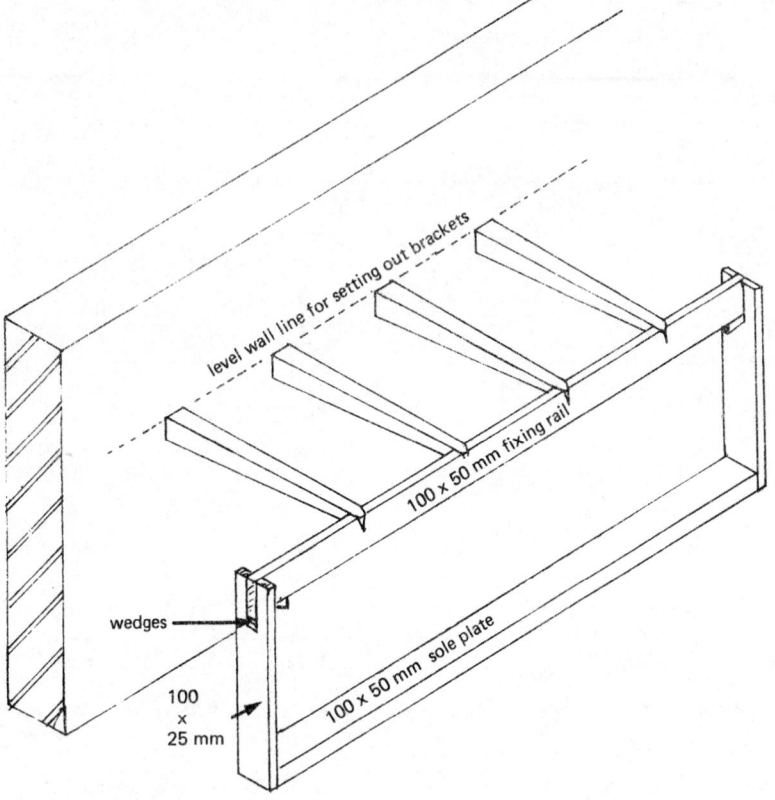

Figure 7.20 Method of fixing a line of cantilever brackets

end of the brackets. After the timber jig has been set up, the brackets should be temporarily inserted and checked for height, level and alignment. When this operation has been completed the brackets should be removed and the holes for the brackets then damped only sufficiently to ensure adhesion.

Mortar for making good around the end of the brackets should be cement and sand, reasonably dense, with a ratio of 1:3 or 1:4, although it should not exceed the density of the walling material. The first bracket is then fixed into position, with the square shank end inserted into the hole formed in the wall face, and around the shank the semi-stiff cement mortar can now be compacted. If the hole is over-large, small particles of bricks can be inserted. Where the walling material is of hollow blocks, then before fixing is carried out, the hollow portions of the blocks should be made solid; this operation should be performed at least 24 hours before the actual fixing of the brackets.

After all the brackets have been fixed, a further check should be carried out for level and alignment. The fixing equipment should not be removed until at least 48 hours after fixing. Care should be taken not

to disturb the brackets, therefore the folding wedges will facilitate the easing operation.

The same procedure can be used for fixing a single bracket, but only a single prop is required to provide support (figure 7.21).

FIXING RAG BOLTS

Rag bolts, often termed holding-down bolts, are normally hand made from wrought iron. They are formed with a ragged base and sides to prevent displacement and ensure complete security in the flooring material. The tops of the bolts are circular, with turned threads to receive washers and a threaded nut (figures 7.22 and 7.23).

Rag bolts are used extensively when heavy, moving machinery is to be anchored to concrete floors. The diameter of the top of the bolt is formed to fit into the hole or slot at the base of the machine.

Setting Out

It is important that considerable care is exercised at the setting-out stage, before the rag bolts are

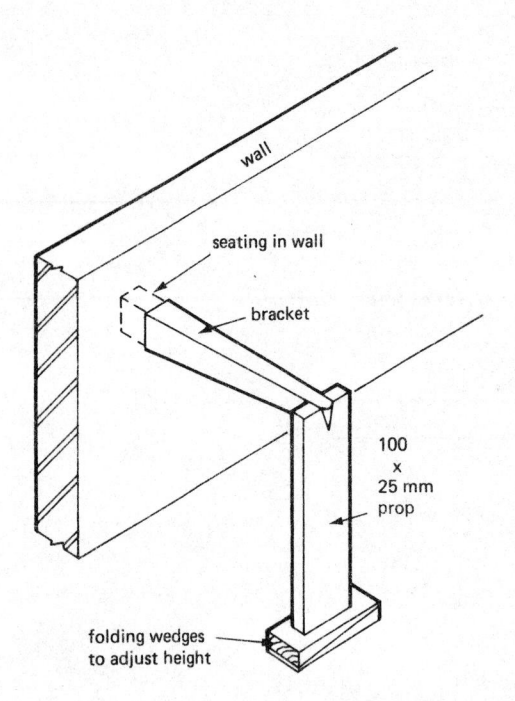

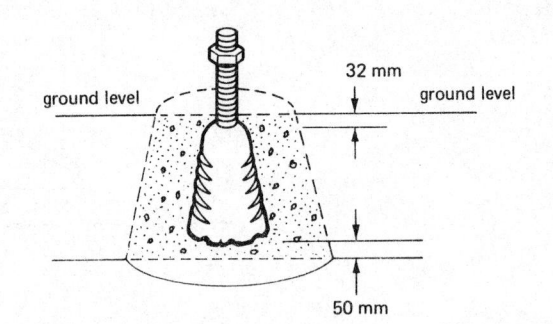

Figure 7.22 Rag bolt with nut and straight base

Figure 7.21 Fixing a single cantilever bracket

Figure 7.23 Rag bolt with nut and ragged base fixed in hole and surrounded with concrete

fixed. Measurements should be checked, levels taken, squareness ascertained and, where there is any doubt, checks should be made again. It cannot be over-stressed that accuracy is of the highest importance since any error could prevent the efficiency and functioning of the machinery. It is obviously very necessary to construct a timber templet, which can be used to aid the setting out and to assist with fixing operations (figure 7.24).

Hole drilling should be carried out with a light-weight Kango, but where the concrete is excessively thick and very dense, it is advisable to use a heavy Kango. The use of the lump hammer and cold-steel chisel, although useful for drilling and cutting a single

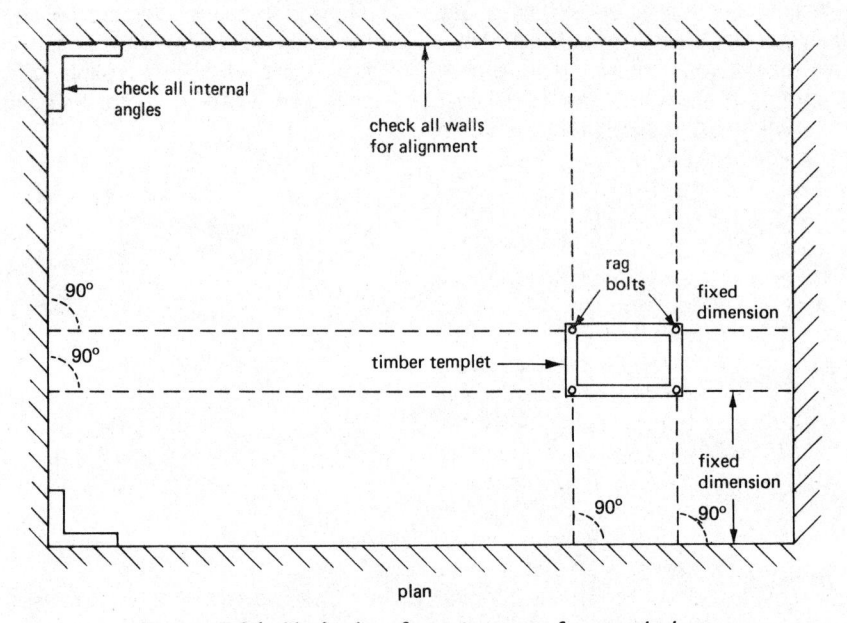

Figure 7.24 Methods of setting out for rag bolts

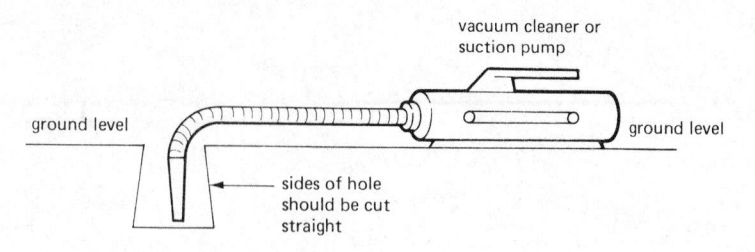

Figure 7.25 Method of cleaning hole after cutting

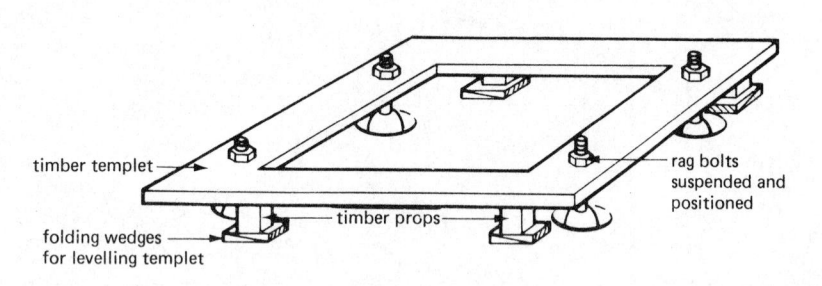

Figure 7.26 Adjusting and positioning rag bolts

hole, would be uneconomical where several holes are required.

After the holes are formed, cleaning out is then required (figure 7.25) and the sides of the concrete should be adequately damped. The timber templet, which positions and supports the bolts, is then placed according to the engineer's drawings (figure 7.26).

The bolts are suspended from the templet in the holes below and a concrete mix of density equal to that of the existing concrete floor is then made. Where the diameter of the holes is less than 100 mm the aggregate size for the concrete should not exceed 12 mm but 18 mm aggregate can be used for larger holes. The water content of the concrete should be reduced to produce a minimum of laitance. The concrete is then placed around the bolt and compacted with a 19 mm-diameter rod and the surface is finished off with the steel float. It is important to cure the concrete around the bolts and a minimum period of 72 hours should be allowed.

The removal of the timber templet should be undertaken with considerable care. The nuts are removed and the templet gently lifted from the bolts. Threads should be checked for cleanliness and where necessary covered with a suitable grease; before leaving, it is again advisable to carry out another check on all the bolts for level, position and dimensions.

8
PAVING

Europeans are extremely fortunate in being able to enjoy and appreciate areas of paving, which have been their heritage for many centuries.

Paved surfaces allow the pedestrian to move with the minimum of fatigue, they complement and enhance buildings and decorative features, and have stood the test of time when used for roads. It is an accepted fact that, in comparison with other forms of construction, the merits of the paved surface are the least recognised. The skills of the designer and craftsman do not receive the appreciation that they deserve. This is possibly because the pedestrian accepts and is familiar with the paved surface. It is to be hoped that this trend is now being reversed and that people are becoming more aware of the aesthetic qualities and importance of the paved surface.

Before considering the type of paved surface that is to be constructed, it is necessary to obtain the following information

(1) the total area of paving required
(2) whether the situation is external or internal
(3) the type of traffic expected
(4) the decorative requirements
(5) the amount of abrasive resistance required for the surface
(6) the requirements for removal of surface water.

When the requirements have been determined, the types of material can then be selected. Paved areas are normally formed by with the following materials: concrete slabs, stone slabs and paving bricks.

TYPES OF PAVING SLAB

Concrete Paving Slabs

These are obtainable in two forms, the pressed, vibrated and reinforced slab and the pressed, unreinforced slab. Concrete slabs can have smooth or non-slip surfaces. They are produced in sizes varying from 600 x 600 x 50 mm to 900 x 600 x 75 mm, although it is possible to obtain smaller slabs of 300 x 300 x 38 mm.

The usual method of cutting unreinforced concrete slabs is with the hammer and chisel or portable electric saw, but the need to cut *reinforced* slabs should always be avoided because failure is almost certain owing to the reinforcement.

Stone Slabs

These are usually made from sandstone or Yorkstone and they are produced with sawn surfaces. The sizes of stone slabs vary, but for normal highway construction 900 x 600 x 75 mm slabs are used. Stone paving slabs are now considerably more expensive than concrete slabs but they are often more resistant to abrasion and possess more decorative qualities. A disadvantage with stone slabs is that they easily become stained, and constantly increasing costs tend to preclude their use, especially for public footpaths; as renewal becomes necessary the concrete slab is usually used as replacement.

Stone slabs are usually cut with the hammer and pitching chisels. Before laying and fixing any type of paving it is necessary to determine the amount of fall or slope required to remove surface water. This is normally between 1 in 40 to 1 in 60, but the amount of fall provided should not increase the physical effort of walking. The direction of traffic is required to determine the direction of the joints in the paving (figure 8.1).

Laying Concrete Slabs

These are usually bedded on a sub-base of sieved clinker ash or sand and placed on mortar dabs under the corners and centre of the slab. The thickness of the mortar dabs forming the bed should not exceed 32 mm and joints between slabs should not exceed 6 mm. The mortar used for bedding is 1:6 cement—sand or 1:2:6 cement—lime—sand (figures 8.1 and 8.2). The same bedding treatment and joints are used for stone slabs, and the mortar mix is also the same.

Concrete and stone slabs should be jointed before the end of the day's work and with the same ratio of mix as for bedding.

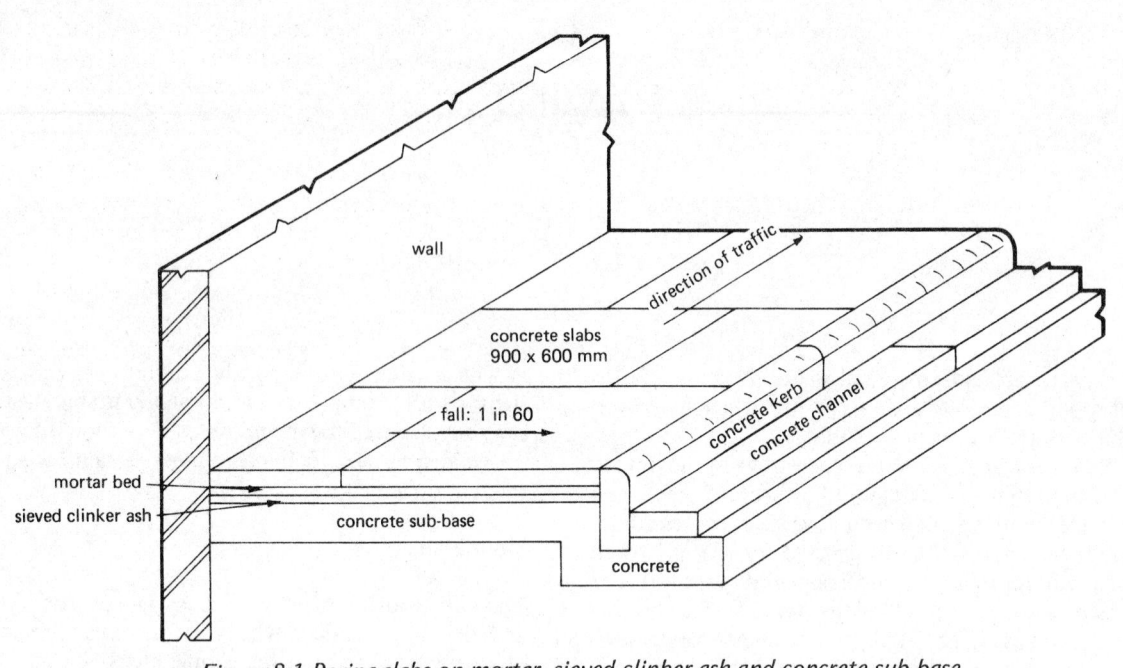

Figure 8.1 Paving slabs on mortar, sieved clinker ash and concrete sub-base

Treatment of Concrete and Stone Slabs

When stacking and storing concrete or stone slabs it is necessary to stack them on edge to ensure complete dryness of the slabs, therefore, whenever possible, these paving materials should be covered with light-weight sheeting.

Tools and Equipment for Laying Paving Slabs

The hardcore base is usually positioned with normal excavating tools — shovels, picks and hammer — but for consolidation a rammer or punner is necessary. These are hand or mechanically operated.

Slabs are usually laid with the beedle or mawl,

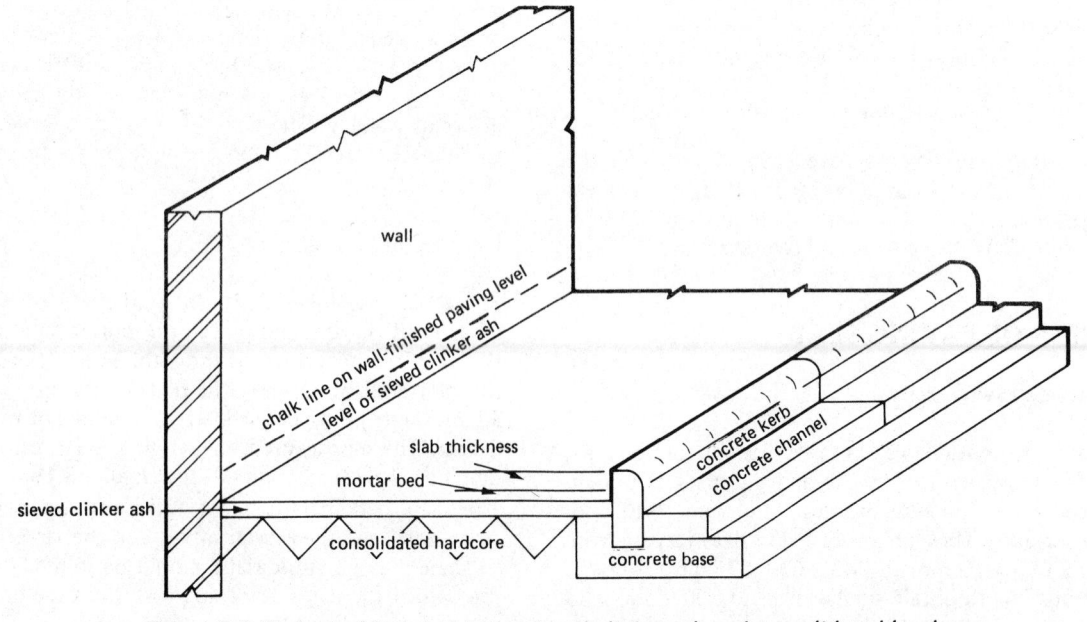

Figure 8.2 Method of laying slabs on sieved clinker ash and consolidated hardcore

which is a large rubber-headed hammer. Paved areas are normally levelled with the aid of the Cowley level and wooden pegs, although boning rods are often used. Straightedges used for checking the slabs are often tapered to the amount of fall required (figure 8.3).

BRICK PAVING

This is the most attractive and decorative form of paving. Because of the great variety of clays found in European countries, a considerable variation in colour and texture can be obtained. The flexibility of bonding arrangements allows the geometrical patterns to enhance the colours of the paving. Brick paving can be used for industrial flooring, where the qualities of the paviors are used to combat abrasive wear and tear, or for domestic use, either external or internal, and also where a decorative appearance is necessary to provide the aesthetic qualities required by the architect.

Brick paving is formed with bricks laid on edge or flat; although the pressed brick is obviously the better type of pavior, wire-cut bricks are often used laid on edge or flatways.

For industrial use engineering bricks or paviors are necessary. The surface of special paviors, produced to withstand very abrasive wear and acids, and also to provide a non-slip surface, is often chequered or impregnated to form a dimpled pattern. These types of brick should be produced to meet the requirements of BS 3679. When cutting is required an abrasive wheel, that is, a brick saw, should be used.

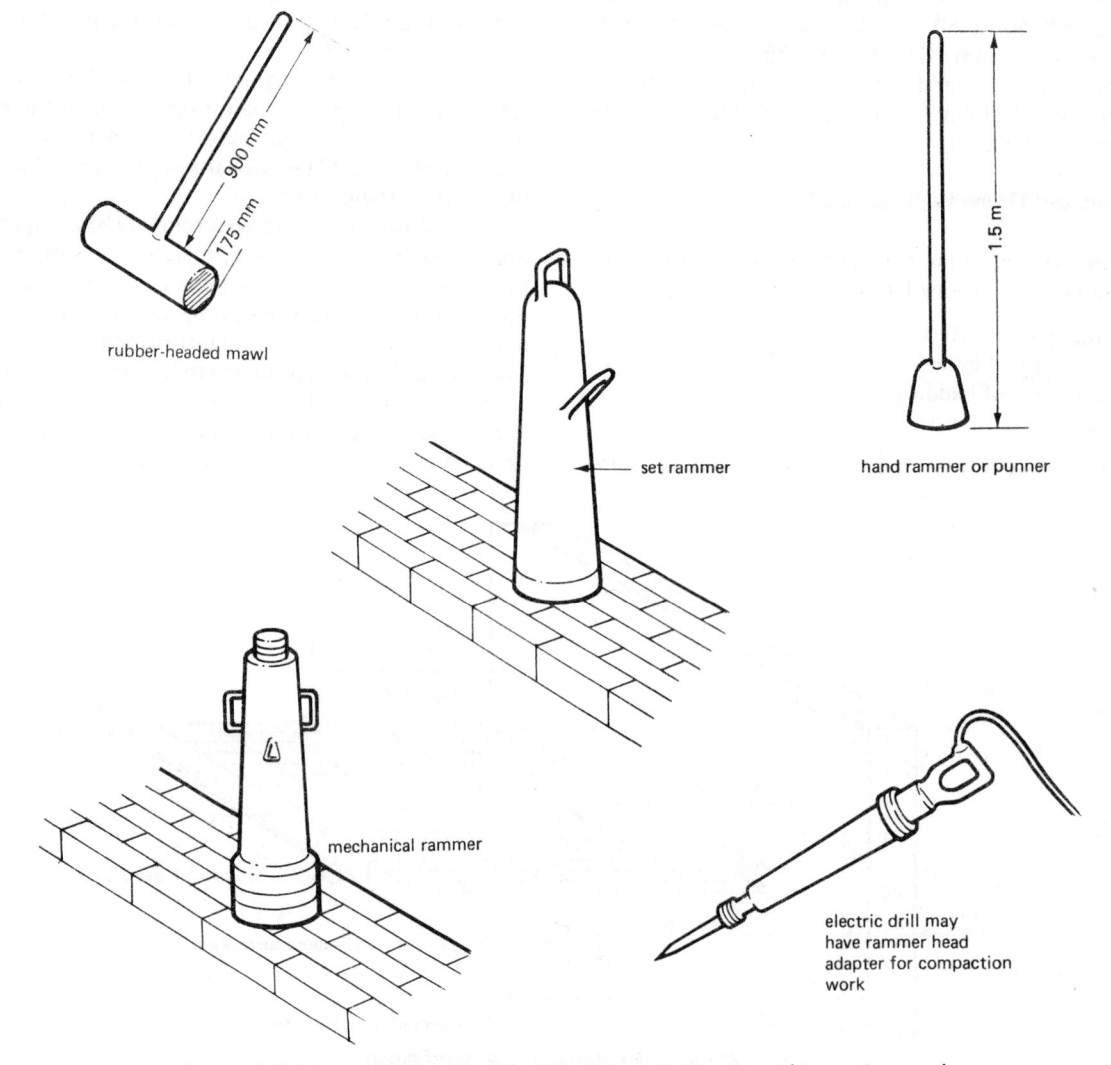

rubber-headed mawl

set rammer

hand rammer or punner

mechanical rammer

electric drill may have rammer head adapter for compaction work

Figure 8.3 Compaction and consolidation equipment used in paving work

Jointing

This should be carried out after the required area has been paved, but during the same day. The mix ratio for the jointing mortar should be the same as for the bedding mortar, but with a minimum water content, and the mortar should be semi-stiff.

Another method of jointing used for paved surfaces is to use a semi-dry mix, and gently brush over the surface of the paving with a fibre brush. The disadvantage of this method is that compaction of the joints may not be complete, also staining of the surface can occur.

Direction of Joints

When designing the paving the direction of the joints should always be considered. When the paving is required for industrial use and to withstand abrasion the paving units should always be bonded and longitudinal joints should be eliminated. The transverse joints should always be at right-angles to the direction of the traffic (figure 8.5).

Movement Joints

These should always be inserted where the sub-base and the paved surface abut any walling, piers, columns or machinery, or when the paved area exceeds 6.0 m in any direction. Obviously, the size of the bay will determine the amount of movement, which will also be influenced by thermal activity and the movement of the paving materials themselves.

To accommodate movement, a joint should be formed around the perimeter between the paving and walling by the insertion of timber battens. These are withdrawn before the curing stage and a movement joint is inserted. The materials used to accommodate movement in paving are butyl rubber, polyurethane and silicone rubbers. Where necessary the surface of the joint can be protected by forming a sealing joint of polysulphide compound (figure 8.9).

Laying Techniques

Paving work should only be performed when other craftsmen have completed their operations. It should be programmed as a finishing operation and only the actual decorative finishing should be performed after the paving work. During the entire operation of paving it is essential for the craftsman to work above the level of the paving, that is, while he is laying the bed, placing and fixing the bricks, and also tamping and jointing. With this method the craftsman is never in a position where he may disturb the bedding or cause any misalignment of the paving (figure 8.8).

Curing Paved Surfaces

After the jointing operation has been completed, the finished work should be allowed to mature. This is assisted by curing. Where it is deemed necessary the floor surface can be protected with light timber battens and polythene sheeting. After a period of 24 hours, the surface can be damped by applying a fine spray of water. This treatment should be continued for a further 72 hours and, for complete maturity, the surface should be closed to all traffic for another 48 hours.

Brick Paving Patterns

Patterns for paved areas are usually determined by the following

(1) the area involved
(2) the surface resistance required
(3) the decorative requirements.

While the same bonding arrangements are obtained with both bricks on edge and bricks laid flat, the decorative appearance in each case may be quite different. The brick on edge contains more joints and yet possibly provides a stronger surface area. The patterns should always be arranged to enhance the area involved. A Flemish or stretcher bond arrange-

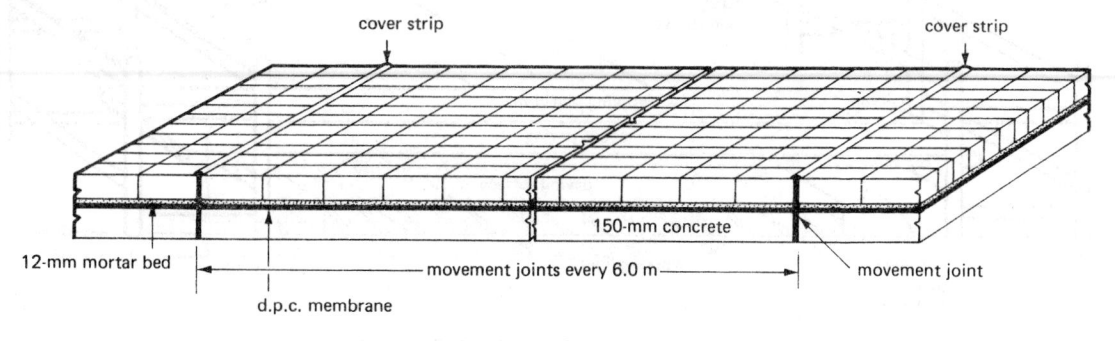

cover strip   cover strip

150-mm concrete

12-mm mortar bed |————— movement joints every 6.0 m —————►| movement joint

d.p.c. membrane

Figure 8.9 Position of movement joints

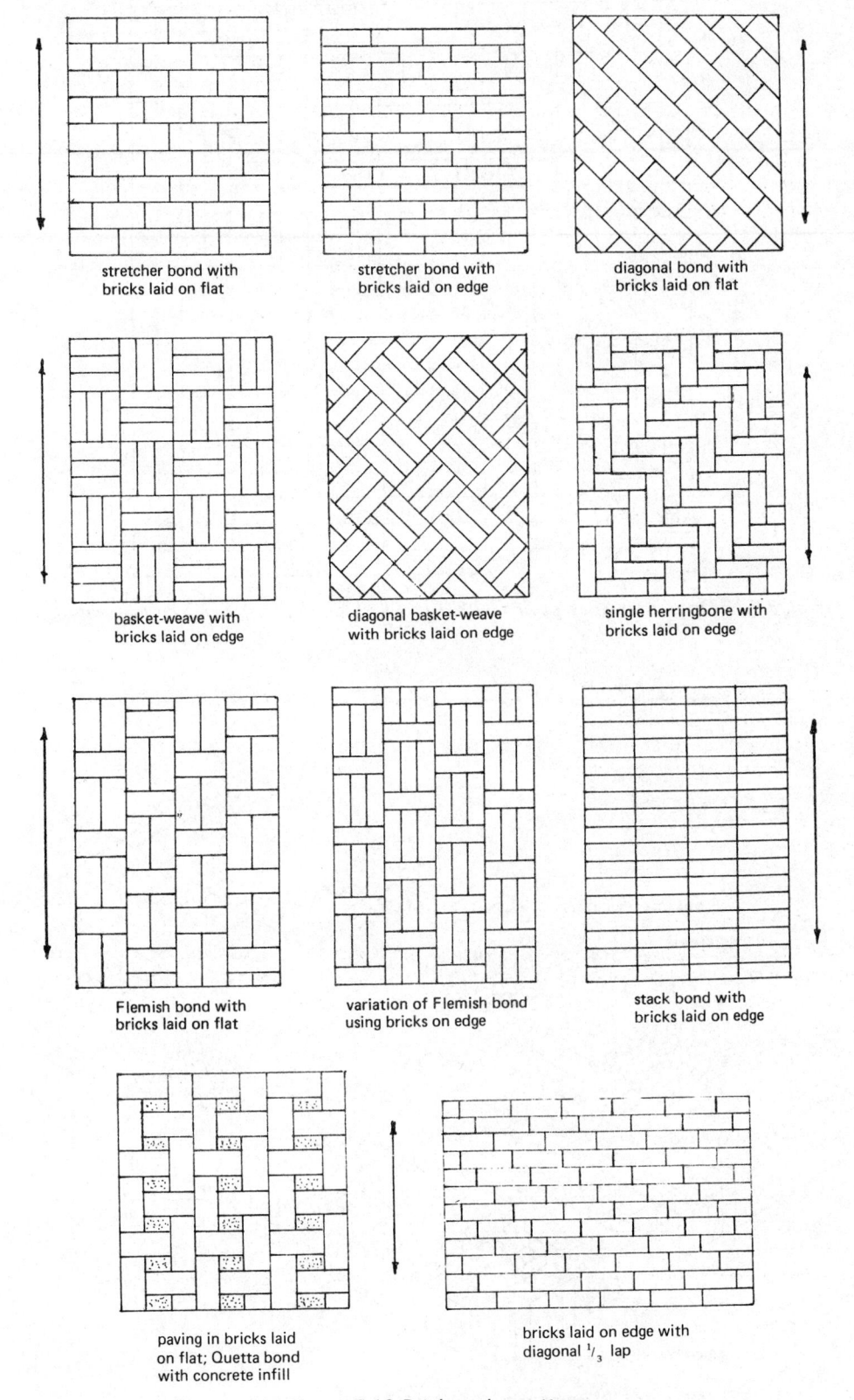

Figure 8.10 Brick paving patterns

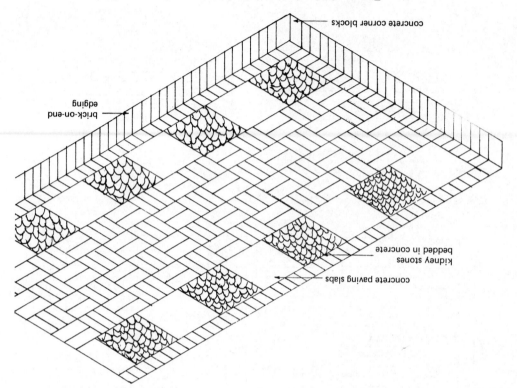

concrete corner blocks

brick-on-end edging

Kidney stones bedded in concrete

concrete paving slabs

Figure 8.12 Compound paving used externally

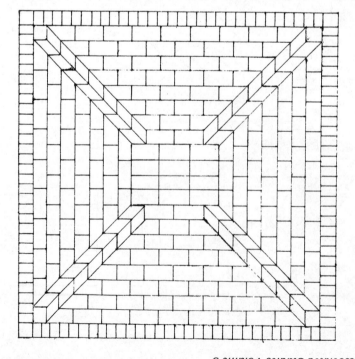

Figure 8.11 Decorative brick paving with bricks laid flat with brick on end to form edging

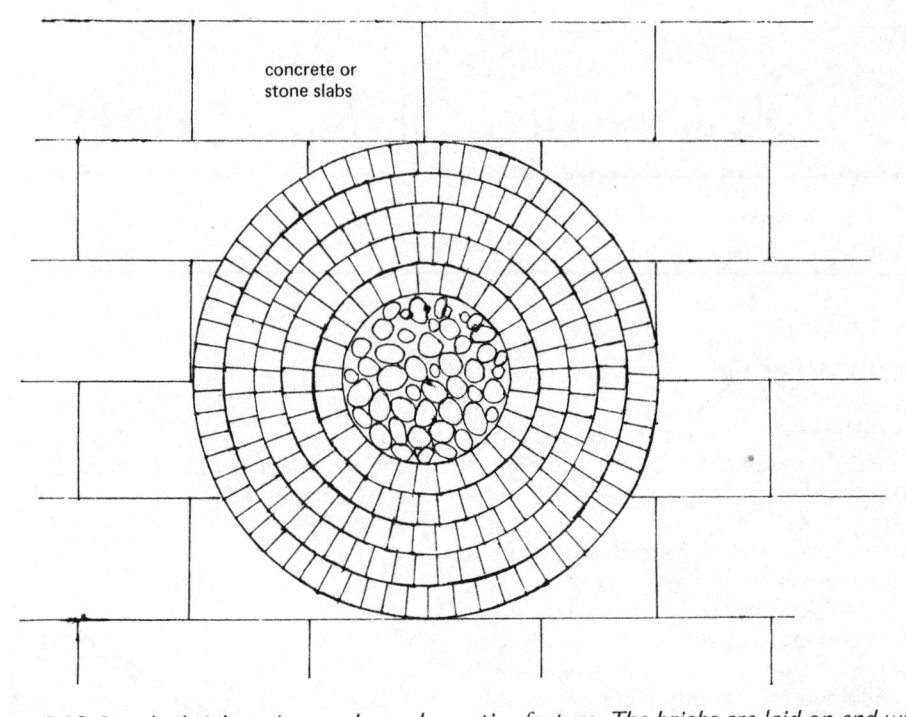

concrete or
stone slabs

*Figure 8.13 Circular brick paving used as a decorative feature. The bricks are laid on end with
kidney stones or exposed aggregate as infill*

ment is often suitable for large areas of paving, while narrow areas are enhanced by a diagonal or herring-bone arrangement. When setting out the bonding pattern it is essential to eliminate the amount of brick cutting, therefore rectangular areas should normally be set out from a base line, and square areas from the centre. Both methods will normally ensure that cut bricks, where required, will be the same size and occur around the perimeter of the area. It may also be advisable to adjust the thickness of joints because on large areas this can eliminate a considerable amount of cutting (figures 8.10 and 8.11).

COMPOUND PAVING

Paved areas formed with more than one material are termed compound paving. Designers are now using combinations of bricks, concrete slabs and often flints and kidney stones (figures 8.12 and 8.13). When the craftsman is required to lay areas of compound paving, it is important that considerable care is exercised at the setting-out stage and when jointing.

It is often the case that a different type of mortar is required for each material used to form the paving. The following recommendations should be kept to when laying compound paved areas.

(1) Always set out the brick paving first.
(2) Check that the dimensions for slabs do not involve considerable cutting.
(3) The brick paving should always be laid first and jointing should be completed before a start is made on laying the other paving materials.
(4) Where the paving is internal, and longitudinal joints separate the different paving materials, movement joints should be inserted between the different materials.
(5) The completed paved area should be matured before traffic is allowed on the paved surface.

When there is a requirement for brick paving to contain pockets of concrete infill, the paving should be completed and pointed before the *in-situ* concrete infill is placed. During placing of the concrete it is good practice to protect the brick paving around each pocket with lightweight plastic sheeting to prevent staining of the brick paving.

9
QUANTITIES OF MATERIALS

BRICKS AND MORTAR

To calculate the number of bricks and the amount of mortar required for any project, the procedure is quite straightforward.

(1) Calculate the area of brickwork, deducting from this figure the area of any openings. The area is found by multiplying the length by the height and it is important to carry out all calculations in metres. For example, if a window opening is given as 600 mm by 600 mm these figures must be multiplied as 0.6 x 0.6 m.
(2) Multiply the area by the number of bricks per square metre. where

thickness of walling	no. of bricks/m^2
½ brick	60
1 brick	120

For example, in 5 m^2 of half-brick-thick walling there are 5 x 60 = 300 bricks.
(3) Add a percentage (usually 5 per cent) to compensate for any damaged bricks, etc. Perhaps the simplest way to find 5 per cent of any number is first to find 10 per cent by moving the point backwards one place and dividing the figure obtained by 2. For example, if 300 bricks are required, then 10 per cent of this is 30 and 5 per cent is 15. Thus the total number of bricks required is 300 + 15 = 315.
(4) While for accuracy mortar should be ordered in cubic metres per square metre of brickwork, it is sufficiently accurate for the craft student to understand that 1 tonne of mortar is enough to lay 1000 bricks, because this very much simplifies the procedure. Since 1 tonne is 1000 kg, it takes 1000 kg to lay 1000 bricks, which is 1 kg per brick. Therefore it will take 315 kg (0.315 tonnes) to lay 315 bricks, and 1.315 tonnes for 1315 bricks, etc.

Example 9.1

Calculate the number of bricks and the amount of mortar required to complete the area of walling shown in figure 9.1.

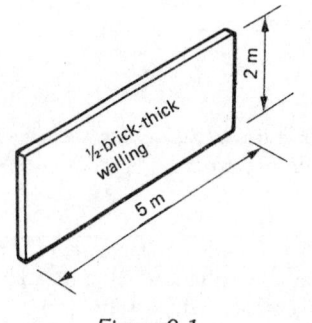

Figure 9.1

$$\text{Area of walling} = \text{length} \times \text{height}$$
$$= 5 \times 2$$
$$= 10 \text{ m}^2$$

$$\text{Number of bricks required} = \text{area} \times \text{no./m}^2$$
$$= 10 \times 60$$
$$= 600$$

Add 5%

$$10\% = 60.0 \text{ (move the point back one place)}$$

Therefore

$$5\% = 30 \quad \text{(dividing by 2)}$$

Total number of bricks
required
$$= 600 + 30$$
$$= 630$$

and

$$\text{amount of mortar} = 630 \text{ kg (0.63 tonnes)}$$

Example 9.2

Calculate the number of bricks and the amount of mortar required to complete the area of walling shown in figure 9.2.

$$\text{Area} = \text{length} \times \text{height}$$

$$= 8.6 \times 2.5$$

$$= 21.5 \text{ m}^2$$

$$
\begin{array}{r}
8.6 \\
2.5 \\
\hline
1720 \\
430 \\
\hline
2150 \\
\end{array}
$$

$$\text{Number of bricks} = \text{area} \times \text{number} /\text{m}^2$$

$$= 21.5 \times 120$$

$$= 2580$$

$$
\begin{array}{r}
21.5 \\
120 \\
\hline
21500 \\
4300 \\
\hline
2580.0 \\
\end{array}
$$

Add 5%

$$10\% = 258$$

Therefore

$$5\% = 129$$

$$\text{Total number of bricks} = 2580 + 129$$

$$= 2709$$

and

amount of mortar = 2.709 tonnes (2 tonnes, 709 kg)

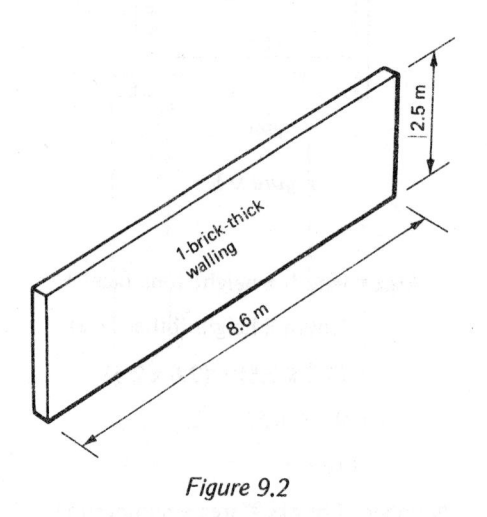

Figure 9.2

Example 9.3

Calculate the number of bricks and the amount of mortar required to complete the walling shown in figure 9.3.

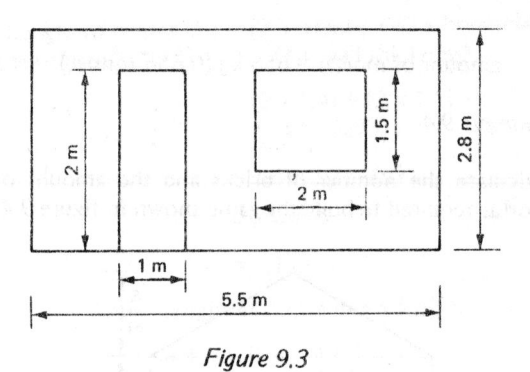

Figure 9.3

$$\text{Area of brickwork} = \text{total area} - \text{area of doors and window}$$

$$\text{total area} = \text{length} \times \text{height}$$

$$= 5.5 \times 2.8$$

$$= 15.4 \text{ m}^2$$

$$
\begin{array}{r}
5.5 \\
2.8 \\
\hline
1100 \\
440 \\
\hline
1540 \\
\end{array}
$$

$$\text{Area of window} = \text{length} \times \text{height}$$

$$= 2 \times 1.5$$

$$= 3 \text{ m}^2$$

$$\text{Area of door} = \text{length} \times \text{height}$$

$$= 1 \times 2$$

$$= 2 \text{ m}^2$$

$$\text{Total area of openings} = 3 + 2$$

$$= 5 \text{ m}^2$$

Therefore

$$\text{area of brickwork} = 15.4 - 5$$

$$= 10.4 \text{ m}^2$$

$$\text{Number of bricks} = \text{area} \times \text{number}/\text{m}^2$$

$$= 10.4 \times 60$$

$$= 624$$

Add 5%

$$10\% = 62.4$$

Therefore

$$5\% = 31.2$$

$$= 32 \text{ (to the number above)}$$

Therefore

$$\text{total number of bricks required} = 624 + 32$$

$$= 656$$

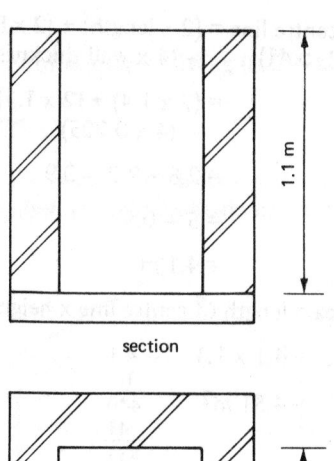

section

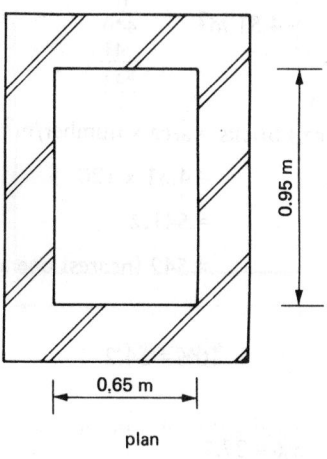

plan

Figure 9.7

examination will reveal that this is exactly the same inspection chamber, only in this case the internal dimensions have been given. Thus calculations from this point are exactly as shown in example 9.6.

ENGLISH AND FLEMISH BONDS

Where walls are built one brick thick and over with English or Flemish bond, the number of bricks required per square metre, assuming a fair face is necessary on one side only is

| English bond | 94 |
| Flemish bond | 84 |

Therefore, to calculate the number of facings and commons required for any area of walling, the procedure is as follows.

(1) Calculate the area of brickwork.
(2) Calculate the total number of bricks required (area x number/m²).

(3) Multiply the area by the number of facings per m².
(4) Deduct the number of facings from the total number of facings to obtain the number of commons.

A simple example of each bond is shown below.

Example 9.8

A wall 6 m long and 1.5 m high is to be built in English bond one brick thick. Calculate the number of facings and commons required.

$$\text{Area} = \text{length} \times \text{height}$$
$$= 6 \times 1.5$$
$$= 9 \text{ m}^2$$
$$\text{Number of bricks} = \text{area} \times \text{number/m}^2$$
$$= 9 \times 120$$
$$= 1080$$
$$\text{Number of facings} = 9 \times 94$$
$$= 846$$
$$\text{Number of commons} = 1080 - 846$$
$$= 234$$

Example 9.9

A wall 5 m long and 2.5 m high is to be built in Flemish bond one brick thick. Calculate the number of facings and commons required.

$$\text{Area} = \text{length} \times \text{height}$$
$$= 5 \times 2.5$$
$$= 12.5 \text{ m}^2$$
$$\text{Number of bricks} = \text{area} \times \text{number/m}^2$$
$$= 12.5 \times 120$$
$$= 1500$$
$$\text{Number of facings} = 12.5 \times 84$$
$$= 1050$$
$$\text{Number of commons} = 1500 - 1050$$
$$= 450$$

BLOCKS AND MORTAR

There are ten 450 x 100 x 215 mm blocks per square metre, which is a very convenient number for use in calculations. As with brickwork the procedure is: find the area, multiply this by the number per square metre, and add 5 per cent for wastage where an allowance is required.

The amount of mortar required for solid block-work 100 mm thick is approximately one-third of the quantity required for brickwork of a similar thickness, that is, since one brick takes 1 kg of mortar and a block is equal in area to six bricks, one block takes 2 kg of mortar and 500 blocks take 1 tonne.

Example 9.10

Calculate the number of 450 x 100 x 215 mm blocks and the amount of mortar required to build figure 9.8 including 5 per cent for wastage.

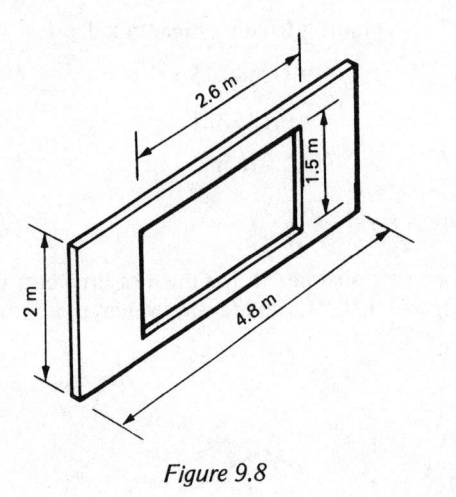

Figure 9.8

Area of blockwork = total area − area of opening

$$\text{Total area} = \text{length} \times \text{height}$$

$$= 4.8 \times 2$$

$$= 9.6 \text{ m}^2$$

$$\text{Area of opening} = \text{length} \times \text{height}$$

$$= 2.6 \times 1.5$$

$$= 3.9 \text{ m}^2$$

$$\text{Area of blockwork} = 9.6 - 3.9$$

$$= 5.7 \text{ m}^2$$

$$\text{Number of blocks} = \text{area} \times \text{number/m}^2$$

$$= 5.1 \times 10$$

$$= 51$$

Add 5%

$$10\% = 5.1$$

Therefore

$$5\% = 2.55 \approx 3$$

Therefore

$$\text{total number of blocks} = 51 + 3$$

$$= 54$$

and amount of mortar = 108 kg (2 kg per block)

Example 9.11

Calculate the number of 100 mm blocks required to build the internal leaf of the gable shown in figure 9.9, including 5 per cent for wastage. Calculate also the amount of mortar.

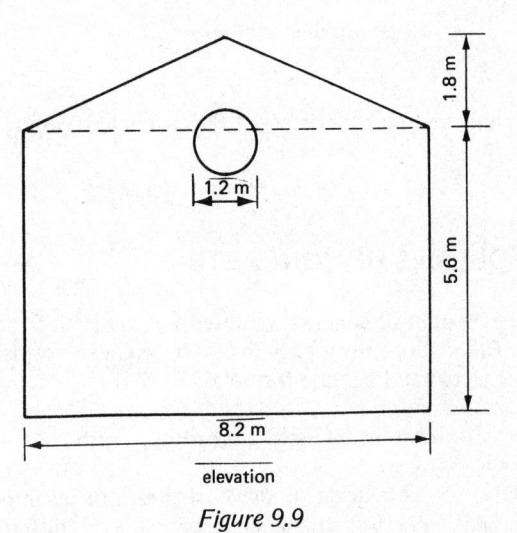

elevation

Figure 9.9

Area of blockwork = total area − area of circular opening

$$\text{Total area} = \text{rectangular area} + \text{triangular area}$$

$$= \text{length} \times \text{height} + \frac{\text{base} \times \text{height}}{2}$$

$$= 8.2 \times 5.6 + \frac{8.2 \times 1.8}{2}$$

$$= 45.92 + 7.38$$

$$= 53.3 \text{ m}^2$$

$$\text{Circular area} = \pi r^2$$

$$= 3.142 \times 0.6 \times 0.6$$

$$= 1.1 \text{ m}^2$$

Area of blockwork = total area − circular area

$$= 53.3 - 1.1$$

$$= 52.2 \text{ m}^2$$

Number of blocks = area x number/m²

$$= 52.2 \times 10$$

$$= 522$$

Add 5%

$$10\% = 52.2$$

Therefore

$$5\% = 26.1$$

$$= 27 \text{ (nearest one above)}$$

Therefore

total number of blocks = 522 + 27

$$= 549$$

and amount of mortar = 1.098 tonnes (1098 kg)

VOLUMES OF CONCRETE

The amount of concrete required for any project such as foundations, lintels, paths, slabs, etc. can normally be calculated from the formula

volume = length x breadth x depth

Only for somewhat unusual shapes, for example, circular columns, triangular sections, is a different formula required. It is important to carry out all calculations, except for very small quantities, in cubic metres (m³) and therefore where the length, breadth or depth is given in millimetres it must be converted to metres before commencement. For example, to multiply 5 m by 100 mm by 50 mm the figures would be 5 x 0.1 x 0.05 m³.

Example 9.12

A strip foundation for a boundary wall is 12.5 m long, 600 mm wide and 150 mm thick. Calculate the volume.

Volume = length x breadth x depth

$$= 12.5 \times 0.6 \times 0.15$$

$$= 7.5 \times 0.15$$

$$= 1.125 \text{ m}^3$$

Example 9.13

A concrete path is 6.4 m long, 950 mm wide and 60 mm thick. Calculate the volume.

Volume = length x breadth x depth

$$= 6.4 \times 0.95 \times 0.06$$

$$= 6.08 \times 0.05$$

$$= 0.3648 \text{ m}^3$$

Example 9.13

A concrete lintel is 1.2 m long, 100 mm wide and 150 mm deep. Calculate the volume.

Volume = length x breadth x depth

$$= 1.2 \times 0.15 \times 0.1$$

$$= 0.18 \times 0.1$$

$$= 0.018 \text{ m}^3$$

Example 9.14

A constructional hearth has the measurements shown in figure 9.10. Calculate the volume of concrete required.

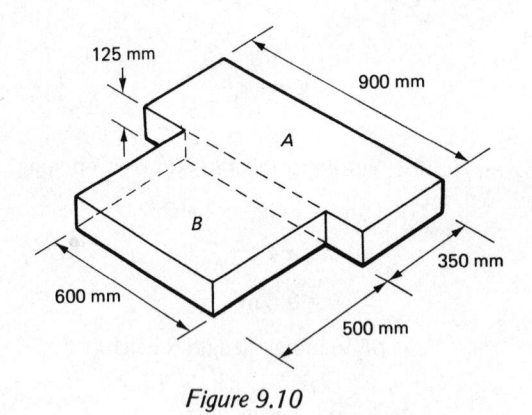

Figure 9.10

Note This must be divided into two parts, as shown by the dotted line. The volume of each part is then calculated separately and the parts are added together to obtain the total volume.

Volume *A* = length x breadth x depth

$$= 0.9 \times 0.35 \times 0.125$$

$$= 0.315 \times 0.125$$

$$= 0.039375 \text{ m}^3$$

Volume B = length x breadth x depth

$$= 0.6 \times 0.5 \times 0.125$$

$$= 0.3 \times 0.125$$

$$= 0.0375 \text{ m}^3$$

Total volume = $A + B$

$$= 0.039375 + 0.0375$$

$$= 0.076875 \text{ m}^3$$

DRY MATERIAL REQUIREMENTS

Concrete

When concrete is to be mixed on site it is sometimes necessary to determine the amounts of cement, fine and coarse aggregate for ordering purposes. While the absolute volume method is the most accurate, because it involves the use of specific gravities it is dealt with in the advanced volume. The following method gives figures slightly in excess of those required.

The density (mass per cubic metre) of well-compacted concrete is approximately 2400 kg/m³, and therefore if the mix is to be 1:2:4 the density should be divided by 7 since there are seven parts $(1 + 2 + 4)$ to give the amount of cement required. This is doubled for the amount of sand and multiplied by 4 for the coarse aggregate. For example

cement = 2400 ÷ 7 = 342 kg (just under 7 bags)

sand = 342 x 2 = 684 kg

stone = 342 x 4 = 1368 kg

Similarly, if the mix is to be 1:3:6 and the density is 2400 kg/m³, the density is divided by 10 since there are 10 parts $(1 + 3 + 6)$ to give the amount of cement. This is multiplied by 3 to obtain the amount of sand and by 6 to obtain the stone. For example

cement = 2400 ÷ 10 = 240 kg (just under 5 bags)

sand = 240 x 3 = 720 kg

stone = 240 x 6 = 1440 kg

Mortar

The density of mortar is approximately 2300 kg/m³ and different mixes are used for different situations. For example, a 1:3 cement/sand is normally used for engineering bricks, whereas 1:6 cement/sand plus plasticiser may be specified for facing brickwork. For aerated concrete blocks used for the inner leaf of

external cavity walls 1:2:8 is quite strong enough. To calculate the requirements for any mix the density should be divided by the number of parts to obtain the cement content and this figure is then multiplied by the number of parts of sand to obtain the sand content.

Example 9.15

Requirements for a 1:3 mix are

$$\text{cement} = \frac{2300}{4} = 575 \text{ kg (11½ bags)}$$

sand   = 575 x 3 = 1725 kg (1.725 tonnes)

Example 9.16

For a 1:6 mix

$$\text{cement} = \frac{2300}{7} = 328½ \text{ kg (about 6½ bags)}$$

sand = 328½ x 6 = 1971 kg (1.971 tonnes)

Example 9.17

For a 1:1:8 mix

$$\text{cement} = \frac{2300}{10} = 230 \text{ kg (just over 4½ bags)}$$

$$\text{lime} = \frac{2300}{10} = 230 \text{ kg (just over 9 bags because a bag of lime contains 25 kg)}$$

sand = 230 x 8 = 1840 kg (1.84 tonnes)

NUMBER OF FLOOR TILES OR PAVING SLABS

To calculate the number of floor tiles or paving slabs required to cover any given area, the procedure is as follows
(1) find the total area to be tiled in square metres
(2) find the area of each tile in square metres
(3) divide the total area to be tiled by the area of each tile
(4) add a percentage for wastage where an allowance is required.

Example 9.18

A room measures as shown in figure 9.11. Calculate the number of 300 x 300 mm tiles required to cover this area.

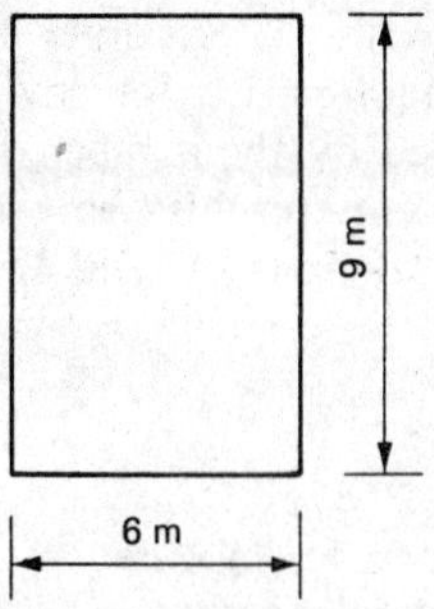

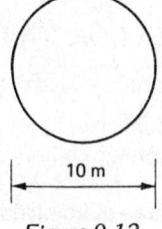

Figure 9.11

Area = length x breadth

= 9 x 6

= 54 m²

Area of each tile = length x breadth

= 0.3 x 0.3

= 0.09 m²

Number of tiles required = $\dfrac{\text{total area}}{\text{area of each tile}}$

$= \dfrac{54}{0.09}$

$\begin{array}{r} 600 \\ 9\overline{)5400} \\ \underline{54} \\ 00 \end{array}$

= 600

Example 9.19

Ignoring wastage, find the number of 500 x 500 mm paving slabs required to pave the circular area shown in figure 9.12.

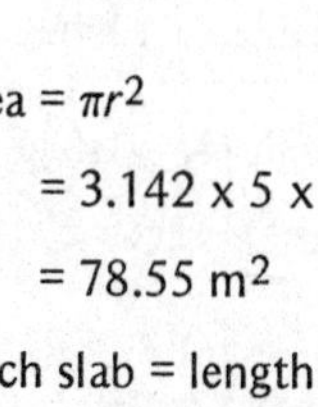

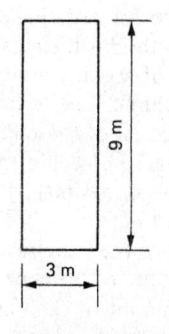

Figure 9.12

Area = πr²

= 3.142 x 5 x 5

= 78.55 m²

Area of each slab = length x breadth

= 0.5 x 0.5

= 0.25 m²

Total number of slabs = $\dfrac{\text{total area}}{\text{area of each slab}}$

$= \dfrac{78.55}{0.25}$

= 314.2

= 315

$\begin{array}{r} 314.2 \\ 25\overline{)7855.0} \\ \underline{75} \\ 35 \\ \underline{25} \\ 105 \\ \underline{100} \\ 50 \\ \underline{50} \end{array}$

Example 9.20

Adding 5 per cent for wastage, calculate the number of 900 x 600 mm paving slabs required to pave the area shown in figure 9.13.

Figure 9.13

Area = length x breadth

= 9 x 3

= 27 m²

Area of each slab = length x breadth

= 0.9 x 0.6

= 0.54 m²

Number of slabs required = $\dfrac{\text{total area}}{\text{area of each slab}}$

$= \dfrac{27}{0.54}$

= 50

Add 5%

10% = 5

5% = 2½

Total number required = 53

Example 9.21

Ignoring wastage, calculate the number of tiles 200 mm square required to complete the floor area shown in figure 9.14.

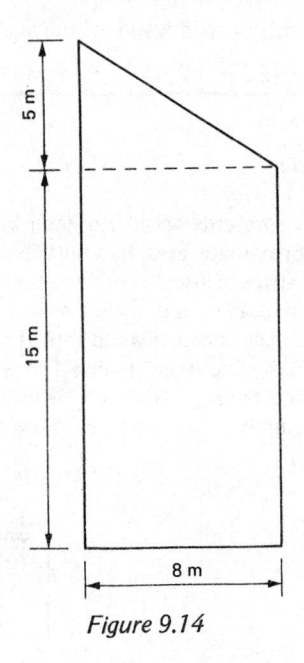

Figure 9.14

Example 9.22

Figure 9.15 shows the plan of a large kitchen floor which is to be tiled with 150 x 150 mm tiles. The area shown shaded is not to be tiled. Calculate the total number of tiles required, making no allowance for wastage.

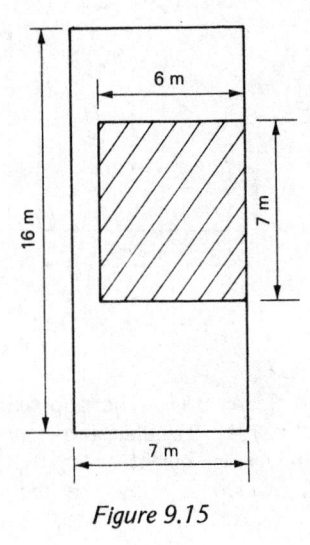

Figure 9.15

Area of rectangle = length x breadth

$$= 15 \times 8$$

$$= 120 \text{ m}^2$$

Area of triangle = $\dfrac{\text{base x height}}{2}$

$$= \dfrac{8 \times 5}{2}$$

$$= 20 \text{ m}^2$$

Total area = 120 + 20

$$= 140 \text{ m}^2$$

Area of each tile = length x breadth

$$= 0.2 \times 0.2$$

$$= 0.04 \text{ m}^2$$

Number of tiles required = $\dfrac{\text{total area}}{\text{area of each tile}}$

$$= \dfrac{140}{0.04}$$

$$= 3500$$

$$\begin{array}{r} 35 \\ 4\overline{)14000} \\ \underline{12} \\ 20 \\ \underline{20} \end{array}$$

Overall area = length x breadth

$$= 16 \times 7$$

$$= 112 \text{ m}^2$$

Shaded area = length x breadth

$$= 7 \times 6$$

$$= 42 \text{ m}^2$$

Total area to be tiled = overall area − shaded area

$$= 112 - 42$$

$$= 70 \text{ m}^2$$

Area of each tile = length x breadth

$$= 0.15 \times 0.15$$

$$= 0.0225 \text{ m}^2$$

Number of tiles required = $\dfrac{\text{tiled area}}{\text{area of each tile}}$

$$= \dfrac{70}{0.0225}$$

$$= 3111$$

	No.	Log
	70	1.8451
	0.0225	$\bar{2}$.3522
	3111	3.4929

Areas of Irregular Figures

The approximate area of an irregular figure may be determined by a number of methods, including the mid-ordinate rule, the squared paper method, the use of measuring instruments, and Simpson's rule. It is usually considered sufficient for craft students to have an understanding of the mid-ordinate rule which is explained as follows.

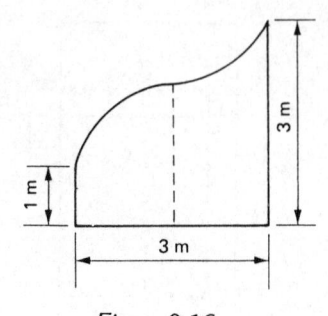

Figure 9.16

Consider figure 9.16. The approximate area of this figure may be calculated by multiplying the width by the average height or length, shown by the broken line, which is known as the mid-ordinate. That is

$$\text{area} = \text{width} \times \text{mid-ordinate}$$
$$= 3 \times 2$$
$$= 6 \text{ m}^2$$

Figure 9.17 shows a larger, more complicated area, and for accuracy this has been divided into a convenient number of strips of equal width. The mid-ordinates are again shown by broken lines.

Area of first strip = width × mid-ordinate
$$= 3 \times 2.5 = 7.5 \text{ m}^2$$

Area of second strip = width × mid-ordinate
$$= 3 \times 3.5 = 10.5 \text{ m}^2$$

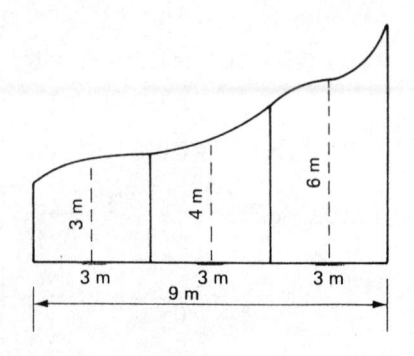

Figure 9.17

Area of third strip = width × mid-ordinate
$$= 3 \times 5 \quad = 15 \text{ m}^2$$

Total area = 7.5 + 10.5 + 15 = 33 m²

A simpler way is to say

Area = width of strip × sum of the mid-ordinates
$$= 3 \times (2.5 + 3.5 + 5)$$
$$= 3 \times 11$$
$$= 33 \text{ m}^2$$

Figure 9.18 represents an area of land 12 m wide. To find the approximate area this must be divided into a number of strips of equal width and the mid-ordinate of each strip drawn and scaled off as shown. The length could have been divided into three 4 m wide strips, four 3 m wide strips, twelve 1 m wide strips or, as shown, six 2 m strips; it should be obvious that the greater the number of strips, the more accurate will be the result.

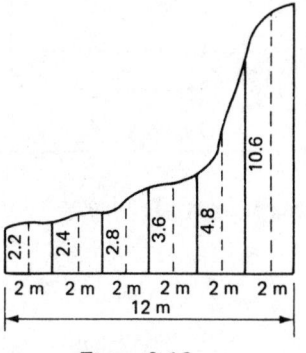

Figure 9.18

To calculate the area using the formula given

Area = width of strip × sum of the mid-ordinates
$$= 2 \times (2.2 + 2.4 + 2.8 + 3.6 + 4.8 + 10.6)$$
$$= 2 \times 26.4$$
$$= 52.8 \text{ m}^2$$

Figure 9.19 shows an area of land 16 m wide which has been divided into four strips, each 4 m in width. The mid-ordinates are drawn and accurately scaled off as before.

Area = width of strip × sum of the mid-ordinates
$$= 4 \times (6.4 + 10.4 + 12.8 + 14.2)$$
$$= 4 \times 43.8$$
$$= 175.2 \text{ m}^2$$

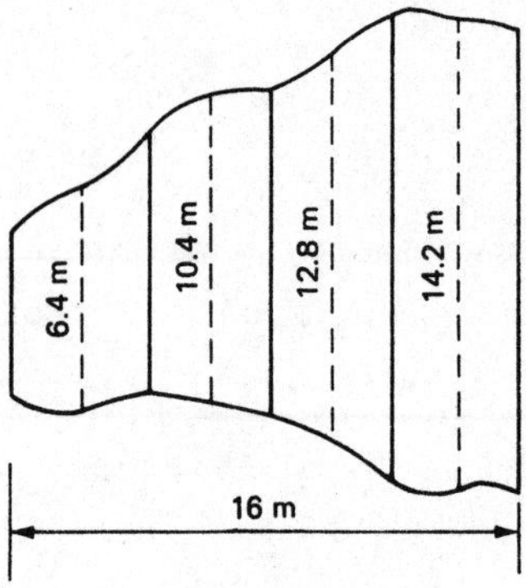

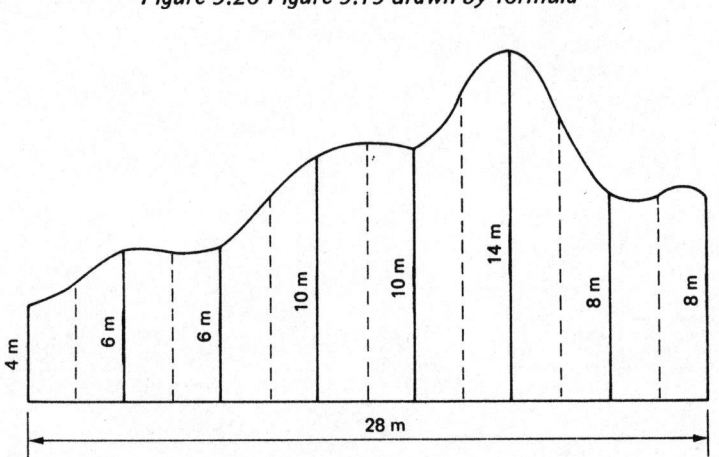

Figure 9.19

Note Figure 9.20 shows why the mid-ordinate rule is fairly accurate; the formula directs that the strips are placed end to end and their overall length is then multiplied by the width of the strip.

One final example is as follows. The irregular figure shown in figure 9.21 is to be covered with a layer of concrete 150 mm in depth. Calculate the volume of concrete required.

Area = width of strip x sum of the mid-ordinates

$$= 4 \times (5 + 6 + 8 + 10 + 12 + 11 + 8)$$
(mid-ordinates averaged)

$$\text{Area} = 4 \times 60$$

$$= 240 \text{ m}^2$$

Volume = area x thickness

$$= 240 \times 0.15$$

$$= 36 \text{ m}^3$$

Note If the ordinates are all given, the mid-ordinates may be averaged (figure 9.21). If not they must be scaled off (figure 9.19).

BRICKS FOR PAVING

When brick-on-edge work is being carried out the number of bricks per square metre is as for face brickwork, that is, 60; but for brick flat paving there are 45 per square metre.

Example 9.23

Calculate the number of paviors required to pave a 8.6 m x 7.5 m courtyard with brick flat paving.

$$\text{Area} = \text{length} \times \text{breadth}$$

$$= 8.6 \times 7.5$$

$$= 64.5 \text{ m}^2$$

Number of bricks = area x number/m²

$$= 64.5 \times 45$$

$$= 2902.5$$

$$= 2903$$

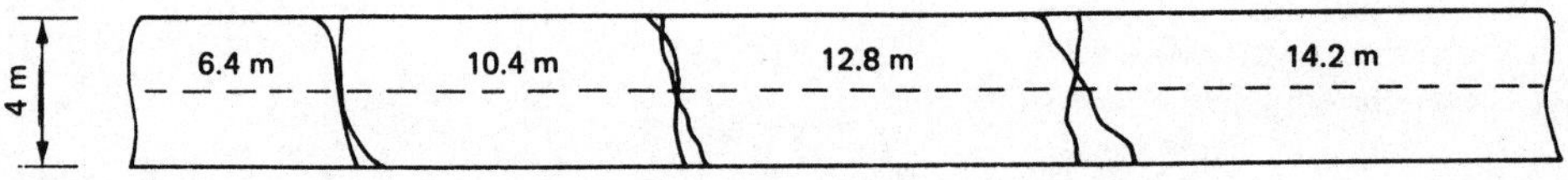

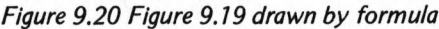

Figure 9.20 Figure 9.19 drawn by formula

Figure 9.21